THE LINUX COMMAND LINE BEGINNER'S GUIDE

JONATHAN MOELLER

DESCRIPTION

The Linux Command Line Beginner's Guide gives users new to Linux an introduction to the command line environment.

In the Guide, you'll learn how to:

-Copy, move, and delete files and directories.

-Create, delete, and manage users.

-Create, delete, and manage groups.

-Use virtual terminals.

-Use the bash shell.

-Safely use the root account with su and sudo.

-Change permissions and ownership of files and directories.

-Create and edit text files from the command line, without using a graphical editor.

-Diagnose network connectivity problems.

-And many other topics.

The Linux Command Line Beginner's Guide

Jonathan Moeller

Originally published in ebook format February 2012.

Created with Vellum

INTRODUCTION

Welcome to THE LINUX COMMAND LINE BEGINNER'S GUIDE! If you're reading this book, you're curious about learning to use the command line interface in the Linux operating system. Fortunately, you've come to the right place. THE LINUX COMMAND LINE BEGINNER'S GUIDE will teach you the underlying concepts and principles of the Linux command line, and show you powerful commands to make the full use of Linux.

WHAT IS LINUX?

What exactly is "Linux", though? Undoubtedly you've used or heard of numerous versions of Linux – Ubuntu Linux, or Linux Mint, or Red Hat Linux. There is a bewildering array of operating systems that call themselves "Linux" - Distrowatch.org, a popular Linux website, tracks the popularity of a hundred different Linux versions on its front page, and there are many hundreds more. With so many different varieties, what exactly constitutes a "Linux" operating system?

"Linux" generally refers to a family of free operating systems based on the Linux kernel (a kernel is the core component of a computer operating system). The Linux kernel is freely available, both to download and modify, and any operating system that runs off the Linux kernel can call itself a Linux operating system. Because it is free, and users have the right to modify it however they please, Linux has been used for a vast array of practical applications.

The history of Linux is long and complex, but we can provide a quick overview here. During the late 1960s and 1970s, AT&T's Bell Labs developed the UNIX operating system. UNIX was a powerful operating system, and universities across the United States soon used it in their computer labs.

However, universities could not modify UNIX to suit their individual needs. AT&T retained rights to the source code, which prohibited any alteration of UNIX. In response to this, computer programmer Richard Stallman launched the GNU Project in 1983. (GNU stands for "GNU's Not UNIX.") Stallman hoped to use the GNU Project to create an operating system similar to UNIX that was nonetheless free to alter and distribute under the principle of "Free Software", a philosophical position which argued that all software should be free to distribute and alter without legal restrictions. The GNU Project and Stallman himself produced a large number of software programs. Unfortunately, the GNU Project lacked a viable kernel, the necessary core of any operating system.

This changed in 1991 when a Finnish university student named Linux Torvald grew frustrated with the academic licensing for Minix, a UNIX-like operating system restricted to educational use only. Torvalds wrote his own kernel, named it Linux, and released it under the GNU free license. Combined with the GNU project, the Linux kernel provided a freely available operating system – an operating system that people could modify, use, and distribute however they saw fit.

Linux had been born.

(Many people argue that the proper name of Linux should in fact be GNU\Linux, in recognition of GNU's vital role, and many GNU programs are used in Linux to this day.)

Because anyone can use or distribute Linux, today Linux and Linux variants run on every conceivable computer platform, from smartphones to desktop computers to high-end server systems. You can run Linux on your personal desktop or laptop computer. It can also power a high-end server farm or a device as small as a smartphone or an MP3 player. Even devices like ereaders, such as Amazon's Kindle or Barnes & Noble's Nook, run using customized versions of Linux.

These different kinds of Linux are generally called Linux "distributions." Anyone, whether an individual or an organization, can create a Linux distribution. Some distributions are commercially supported endeavors, like Red Hat Linux or SuSE Linux. Others, like Knoppix or Fedora, are free and community-supported. Some, like Ubuntu, fall in the middle, and are offered free to the public while supported by an organization.

If you need to accomplish a computing task, odds are that there is a distribution of Linux that will achieve that task.

WHAT IS THE COMMAND LINE?

This is a book about the Linux command line. But what, exactly, is the command line?

The term "command line" is generally used as shorthand for more accurate term, the "command-line interface" (commonly abbreviated as the CLI). Most modern computer systems use what is called a graphical user interface (GUI), whether it is a computer with icons and a mouse, or the touch-based interface of a tablet computer like the iPad. A command-line interface, on the other hands, eschews graphical elements entirely. With a command-line interface, instead of graphical icons and windows, the user only sees a text prompt at which he can enter commands. To launch programs, the user must type the appropriate commands into the prompt and hit the ENTER key.

Most Linux distributions intended for end users to run on home computers use some sort of GUI, typically the GNOME or KDE desktop environments. Linux distributions intended for server use or as network appliances almost always boot to the command line interface. Yet even in graphical Linux environments, you can easily get to the command-line interface via a terminal emulator, by a virtual console, or by simply booting the computer to the command-line interface. (And if the GUI happens to break, a familiarity with the command line will come in handy!)

Knowledge of the Linux command line, therefore, will be valuable no matter what version of Linux you use.

THE PURPOSE OF THIS BOOK

This book is intended to give the reader a thorough grasp of the basics of the Linux command line. It's not designed as an all-inclusive study, but offers enough information to get you started with the basics, and hopefully enough confidence for you to start experimenting on your own. Initially, I envisioned this book as a reference guide, with common Linux commands sorted alphabetically, but I quickly realized this approach would not work. For example, there are numerous commands that deal with user accounts, and for every single command, I would have to explain the underlying concepts of Linux user accounts over and over again.

Instead, I decided to divide this book up into "concept" and "commands" chapters. The "concept" chapters explain some of the underlying concepts of the Linux command line – the file system, for instance, or the way user accounts work. Immediately after each "concept" chapter comes one or more "commands" chapter, describing the commands dealing with that concept. The chapter on the filesystem is followed by a chapter describing the commands to move, copy, and delete files and folders. Hopefully, this approach will give the reader an understanding of the underlying concepts of the command line, as well as a practical knowledge of the related Linux commands.

WHY I WROTE THIS BOOK

I began writing about computers and technology almost by accident. I started my writing career as a writer of fantasy fiction, specifically of sword-and-sorcery novels. (All my books are now available in all major ebook formats; see the "Other Books By The Author" section for links.) Like every good fantasy writer, I had a blog where I attempted to promote my books, and as you might expect, the blog managed to pull down only thirty or forty hits a month.

One day I happened to write a post about I problem I had with the Samba file-sharing server on Ubuntu Linux 8.04. To my very great surprise, the next day I saw that post had received nearly sixty hits via people doing Google searches for Samba problems on Ubuntu. From there it was a short step to blogging regularly about technology (and turning a profit via web ads), and then when I began using a Kindle in 2010, to writing short ebooks on technical topics.

I decided to write this book because I have seen many people attempt to learn Linux, only to give up in frustration with the command line. In one of my previous books, THE UBUNTU BEGINNER'S GUIDE, I devoted several chapters to the command line, but that book focused mainly on setting up Ubuntu as a server. With this book, I hope to help the new user to learn the Linux command-line interface, whether a home user experimenting with Linux or a IT professional seeking to learn new skills.

A NOTE ON LINUX VERSIONS

As mentioned above, there are numerous different Linux distributions. Most of the commands described in this book will work on the majority of different Linux distributions. That said, sometimes individual Linux distributions will configure the commands in a specific way, so sometimes a command or option that works on one Linux distribution will not work on a different distribution. This tends to happen only with the more exotic or esoteric commands and utilities, but it does happen. It is always a good idea to check a distribution's documentation and man pages for precise information (and we'll show you how to do just that in the final chapter of this book).

The screenshots for this book were taken on Ubuntu 11.10, Linux Mint 12 and Fedora 16.

1

CONCEPT: TERMINALS AND VIRTUAL CONSOLES

Before you do anything from the command line, you have to first actually access the command line interface. When you boot up Ubuntu or Linux Mint or another Linux distribution with a graphical interface, you might wonder how to even get to the command line. In this chapter, we'll discuss how to reach the command-line interface both from a GUI-based Linux distribution and a CLI-based one. We'll also show you how to have multiple command-line windows open at once, whether in the GUI or in a pure CLI environment.

ACCESSING THE COMMAND LINE FROM A CLI-BASED DISTRIBUTION

Some Linux distributions, such as Ubuntu Server, do not include a GUI. For such distributions, getting to the command line isn't hard. You simply boot up the computer or server, and the system goes to the logon prompt. Enter your username and password, and you will log into the system and arrive at the command line prompt, also known as a terminal session. The term "terminal" dates back to the days when UNIX ran on large mainframes accessed via remote terminals.

Sometimes you might see a blank screen on a CLI Linux machine that's powered up. This is no cause for panic – to save power, the system is usually configured to power off the screen after a set interval of inactivity. Simply tap the space bar to awaken the display once more.

ACCESSING THE COMMAND LINE FROM A GUI-BASED DISTRIBUTION

Accessing the command line from a GUI-based Linux distribution is a little trickier. By default, a GUI-based distribution boots into the graphical interface. Even within the GUI, it's simple to access a command prompt. There are two common ways of doing so.

The first and most common is an application called a "terminal emulator." As the name implies, a terminal emulator recreates the command line environment, but within a window of your GUI. You can access the full range of Linux commands from the terminal emulator application, but you can still use the benefits of the GUI - the mouse, the menu bar, and copying & pasting. (Copy & paste is particularly useful, since you can copy and paste complex and lengthy commands into the terminal emulator without the hassle of typing them.)

Almost all desktop Linux distributions include a terminal emulator application. You can usually launch it by going to the Applications menu, then to the Accessories category, and finally by clicking on the Terminal icon. If your distribution comes with a quick-launch bar (like Ubuntu's Unity Launcher), very often the icon for the terminal emulator is pinned to the launcher.

The second way to access the command line in a Linux GUI is through what's called a "virtual console." We'll discuss virtual consoles in this chapter, but first we'll describe how to open multiple terminal windows.

OPENING MULTIPLE TERMINAL WINDOWS IN THE GUI

ONE NICE FEATURE of working with the terminal emulator application in the GUI is the ability to open multiple terminal windows. There are several advantages to running multiple terminal windows – if you run a command that displays real-time output (like the top monitoring command), you can open a new window to execute additional commands. Or if you're running a command that take a long time to process, like copying several gigabytes of files, you can open a new window and run another command while the first one finishes working.

Most terminal emulator applications support opening a new terminal window. To do so, you usually go to the File menu and select either "New Window" or "New Terminal." A new terminal window will open up.

If you don't want to clutter up your desktop with a bunch of windows, most terminal applications now offer tabs, much as the Mozilla Firefox and Google Chrome web browsers allow you to open new tabs for different web pages. If you open new tabs in the terminal emulator, a row of tabs will appear below the menu bar, each containing a separate terminal session. Usually, you can open new tabs by going to the File menu and clicking on "New Tab."

Obviously a graphical terminal application will not work in the command-line environment. It is nonetheless possible to run multiple terminal windows from the command line using "virtual consoles", which we will discuss in the next session.

VIRTUAL CONSOLES

A VIRTUAL CONSOLE IS, essentially, a second connection to the same computer. Remember that Linux was based upon UNIX, and most early UNIX machines were massive mainframe computers. To connect to those mainframes, users employed remote consoles or dumb terminals. Because UNIX was a multi-user operating system, multiple users could log into the mainframe at the same time from different consoles.

Virtual consoles are a legacy of those mainframes. Most Linux versions support a maximum of seven "virtual" consoles, software consoles built into the operating system that mimic the function of those old dumb terminals. Basically, a user can log into the operating system using one of those virtual consoles, and then a second user could switch to a different virtual console and log in, leaving the first user's session unaffected (though they would, of course, have to share the keyboard and mouse).

As we mentioned, Linux supports seven virtual consoles. To switch between virtual consoles, press the CTRL-ALT-FUNCTION# keys simultaneously. So to switch to virtual console one, you would hit CTRL-ALT-F1, to switch to virtual console 2, CTRL-ALT-F2, and so forth. When you switch to a new virtual console, you'll see a standard prompt to log in. You (or another user) can then log into the virtual console without affecting your original session.

If you have a version of Linux running a GUI, you can also use virtual consoles. By default, the GUI runs in virtual console seven (CTRL-ALT-F7). You can switch to any of the other six virtual consoles. When you are finished, you can return to the GUI by hitting CTRL-ALT-F7.

2

CONCEPT: THE BASH SHELL

When you use the Linux command line, you are almost certainly using a program called the bash shell. After you've finished reading this chapter, you'll understand what a shell is, and how it relates to the Linux command line. You'll also knew some tips and tricks to wring maximum functionality out of the bash shell.

WHAT IS A SHELL?

IN COMPUTING, A "SHELL" doesn't refer to the hard covering of a mollusk, a kind of pasta, or a type of artillery ammunition. Instead, the term "shell" generally refers to the program that serves as the interface between the user and the computer's operating system. When you use a computer, you are not communicating directly with the operating system. Instead, you are issuing commands through a shell, which then passes the instructions on to the operating system for you.

Both a graphical user interface and a command line interface qualify as a shell. There are numerous different command-line shells available for Linux – the sh shell, the zsh shell, the c shell, and others. The most common shell by far is the bash shell, which we will discuss in the next section.

THE BASH SHELL

PROGRAMMER BRIAN FOX wrote the first version of the bash shell in the later 1980s as part of the GNU Project. The "bash" name is a pun – it stands for Bourne-again shell. The Bourne shell was one of the most popular UNIX shells, but as it was not free software, users could not modify and distribute it legally. The GNU Project and the Free Software Foundation both wanted to create a free alternative to the Bourne shell – hence, the "Bourne-again shell", the bash shell.

The bash shell is far and away the most common Linux shell. Most Linux distributions use the bash shell as their default command shell, and even Mac OS X (which is technically a version of UNIX) uses the bash shell as its default terminal emulator. Familiarizing yourself with the bash shell will prove useful across a wide variety of Linux distributions, and even on Macs.

In the next sections, we'll go over some of the bash shell's useful features.

THE COMMAND PROMPT

FIRST, we will examine the command prompt itself. If you boot into a CLI environment, or you launch a terminal emulator application, the first thing you see is the prompt. (The name comes because it is "prompting" you to enter a command.) The prompt itself might seem like a string on nonsensical words and symbols, but it conveys quite a lot of useful information.

Usually, the prompt looks like this:

jmoeller@yourlinuxmachine:~$

Let's examine what each part of the prompt means.

The portion of the prompt before the @ symbol (in this case, "jmoeller"), is the username of the currently logged-in user. Naturally, this will change based upon which user is logged in to the terminal session. The portion after the @ symbol but before the colon (in this case, "yourlinuxmachine") is the name of the machine you are using. From this prompt, you can see that you are using the account "jmoeller" on the machine named "yourlinuxmachine."

The next part of the prompt, after the colon mark but before the dollar sign, shows the working directory. In the bash shell, the "working directory" is the currently selected directory. The tilde symbol (~) might not seem informative, but in the bash shell, it is a shorthand character for the home directory of the currently logged in user. In the prompt in our example, the tilde character would represent the home directory of jmoeller, /home/jmoeller. As you change the working directory, the prompt changes to reflect that. If you changed the current directory to /etc, the prompt would look like this:

jmoeller@yourlinuxmachine:/etc$

The dollar sign indicates the end of the prompt. Your commands will appear after the dollar sign character.

COMMAND/FILENAME COMPLETION

While you can perform certain tasks more efficiently and quickly from the command line than from the GUI, it does involve a lot of typing, which can become tedious. But using a feature called command/filename completion can cut down on quite a bit of that repetitive typing. The way command completion works, you type out part of the file name and then hit the TAB key. The bash shell then takes it best guess at filling out the rest of the filename for you. Usually, it's fairly accurate.

For example, let's say you wanted to use the rm command to delete a file name Caina1.doc. Type out the command and the first part of the filename:

rm Ca

Hit the TAB key, and the bash shell will finish out the filename for you:

rm Caina1.doc

What happens if multiple files in the same directory have similar names, like Caina1.doc and Caina2.doc? In that case, command/filename completion will sort through the files in alphabetical order. Say you typed this command and then hit the TAB key:

rm Caina

The bash shell would first produce this:

rm Caina1.doc

But if you hit the TAB key for a second time, you would see this:

rm Caina2.doc

You can also use command/filename completion to fill out long directory names. Say you wanted to delete a file called /var/www/storage/documents/Caina1.doc. Typing that out would prove cumbersome. But if you start out by typing this:

rm /var/www/s

Then hit TAB, and you'll see this:

rm /var/www/storage/

Append a "d" to the end of the command:

rm /var/www/storage/d

Hit TAB again, and the bash shell should figure out the directory:

rm /var/www/storage/documents

Finally, add the first letter of the file name:

rm /var/www/storage/documents/C

And one more hit of the TAB key should finish it off:

rm /var/www/storage/documents/Caina1.doc

With command/filename completion, you can type long filenames far quicker than you could otherwise.

COMMAND HISTORY

As you work on the Linux command line, you might find yourself reusing the same commands or similar commands over and over again. You could of course retype them, but the bash shell has a feature called "command history" that makes the retyping unnecessary. Command history recalls the commands you type – usually, it can recall the last two thousand commands. You can recall them to the prompt by pressing the UP arrow key on your keyboard. Keep pressing UP to scroll through the list of commands until you find the one you need. If you go too far into the list, you can use the DOWN arrow key to scroll backward through the commands.

You can also view your entire command history with the history command at the prompt:

history

The actual command history is stored in a configuration file in your home directory, and we will discuss those configuration files in the next section.

THE BASH CONFIGURATION FILES

The bash shell's behavior is governed by a number of hidden files located in your home directory. (In Linux, files are "hidden" when their names are prefixed with a period character. While this does not protect them from deletion or modification, they don't show up in regular directory listings.) For a new user, it's generally best to avoid modifying these files. Once you have more experience, you might want to experiment with modifying them.

The first file is ~/.bash_history. This is where bash stores the information used in the command history feature. It is a plain text file, so you can view it with a number of commands we'll discuss in Chapter 16.

The second file is ~/.bash_profile. This stores the individual settings for your bash settings, including the configuration of the prompt text itself. If you want to customize your bash prompt, you can do so here.

OUTPUT PIPING AND REDIRECTS

THE BASH SHELL also offers two features called "output piping" and "output redirects." Basically, this lets you take the output from a command and either hand it over to another command for processing, or to store it in a text file. This can become quite complicated, but we'll look at a few common uses.

Piping allows you to take the output from one command and move it to a second command for processing. For example, let's say you wanted to use the ls command to view the contents of the current working directory. Furthermore, you want to use the grep text-sorting command to find every file named "cheese" in the directory. To use command piping, you would use the pipe character (|) to pipe the output from the ls command to the grep command:

ls | grep cheese

Output redirection allows you to use the greater than sign (>) to redirect the output from a command to a different location, usually a text file. For example, let's say you wanted to use the ls command to look through the contents of the working directory, but you wanted to save the command's output to peruse later. You can do that by using the ls command with a greater than sign to redirect the output:

ls > list

This will redirect the command's output to a plain text file named list. If the file doesn't exist, the command will create it on the spot in the current working directory. However, if the file does exist, the command will overwrite it completely with the new information. If you're not careful, this can result in data loss.

This data loss is easy to avoid. If you use a double greater than sign (>>), then the output from the command will be appended to the existing file without overwriting it. To expand on our previous example:

ls >> list

This version of the command will append the output from the ls command to the list file, keeping any data from getting overwritten.

FILE NAMES WITH SPACES

THE COMMAND LINE does not handle spaces in file names gracefully – let's say you tried to use the rm command to delete a file named "Joe's Term Paper."

rm Joe's Term Paper

This command would fail because the command line would interpret each word in the command as a separate file. To force the command line to view the file properly, enclose any filename with spaces within quotation marks:

rm "Joe's Term Paper"

This time rm will delete the file.

CASE SENSITIVITY

WE WILL CLOSE this chapter with an important reminder: always remember that the Linux command line interface is case sensitive. That means the bash shell distinguishes between uppercase and lowercase characters. It will see three files named Caina.doc, caina.doc, and CAINA.doc as three different files. In addition, almost all Linux commands are in lowercase. If you intend to type ls to view the contents of the current directory, but accidentally type Ls instead, then the shell will not recognize the command and return an error message.

This often confuses new Linux users, especially ones with backgrounds in DOS or Microsoft Windows. Neither DOS nor the Command Prompt application in Windows are case sensitive – unlike Linux, the Windows Command Prompt application would see Caina.doc, caina.doc, and CAINA.doc as the same file, and you can also type Windows commands in either uppercase or lowercase.

This is not true with Linux. Always remember that the Linux command line interface is case sensitive.

3

CONCEPT: THE LINUX FILESYSTEM

To effectively use the Linux command line, you must first have a strong grasp of the underlying Linux file system. Without knowing how to copy, move, delete, and create files and directories from the command line, the Linux command line interface won't be very useful. In this chapter, we'll explain how the Linux filesystem works. In the chapter after that, we'll demonstrate the commands you need to manage files and directories.

WHAT IS THE LINUX FILESYSTEM?

IF YOU'RE OLD ENOUGH, you might remember when the term "filesystem" meant how an office organized its collection of manilla paper folders. For computers, a "filesystem" refers to how the information stored on a disk is organized, whether on a hard disk, a USB flash drive, or an optical disk like a CD-ROM or a DVD-ROM. (Or a floppy disk, if you're unfortunate enough to have a computer that old.) In Linux, when people talk about the "filesystem", they generally mean the file and directory structure on the main hard disk of the Linux system. In fact, in the most recent versions of Ubuntu Linux, clicking on the "File System" icon in the Nautilus file explorer shell will take you to the top directory of the main hard drive.

The Linux filesystem is divided into files and directories, which we will discuss in the next section.

FILES AND DIRECTORIES

IF YOU'VE USED a computer for any length of time, you're already familiar with the term "file." A computer file is, basically, a piece of information saved to long-term storage in a disk. Your documents, pictures, and music tracks are all files, and so are all the programs and applications you run on your computer (an elaborate application can sometimes consist of hundreds of even thousands of files). Interestingly, in the Linux filesystem, various hardware devices are also represented as files, all stored in the /dev directory.

A modern Linux system contain tens of thousands of files, and keeping those files organized can prove quite a challenge. Fifty thousand files in a single location would make it extremely difficult for either the operating system or the user to find anything. This is where the concept of a directory comes in handy. Directories allow you to separate files into separate virtual containers. Just as as paper folders once allowed office workers to sort a single enormous pile of paper into a neatly organized system, so to do directories allow you (and the operating system) to keep track of files.

(Incidentally, the term "directory" and "folder" mean the same thing. So when you hear someone talk about accessing their Documents folder, they're talking about the same thing as their Documents directory. Since most modern GUIs use a folder icon to represent directories, using the word folder to refer to a directory has become common practice.)

You can, of course, put directories within other directories. These directories-in-directories are generally called subdirectories, and they too can have subdirectories of their own.

You can best visualize the structure of the filesystem by using the tree command:

tree

This will display the filesystem as a tree, starting with the topmost directory, and working its way down the various subdirectories.

Linux's standard subdirectories have been pretty well defined for many years, but we'll start by discussing the most important directory in the filesystem.

THE ROOT DIRECTORY

Directories can contain subdirectories, and those subdirectories can hold additional subdirectories. These directories form an elaborate, hierarchical structure, and the starting point of the that structure is the root directory. The root directory is the topmost directory in the structure, the first directory in the entire file system. Every other directory in the system is technically a subdirectory of the root directory, no matter how many levels deep you have to go (and in complex Linux installations, you can have subdirectories nested a dozen levels deep or more).

In the command line, the root directory is represented with the forward slash (/) symbol. So if you see a location in the file system, like /home/jmoeller, the root directory is represented by the forward slash on the left. That means the home directory is located within the root directory, and the jmoeller directory is in turn stored within the home directory.

It's usually a good idea to keep your root directory as uncluttered as possible. This makes it easier to locate files in the filesystem, and also prevents potential security breaches from unauthorized software running in the root directory. (In the early days of Windows, lazy users had a habit of storing everything in the root directory of their hard drive, which often caused all kinds of problems.)

Regardless of the Linux distribution you use, you will probably see the same subdirectories in the root directory of your filesystem. The Linux directory structure has been standardized for years, and in the next section, we'll take a look at a few of the more common subdirectories.

COMMON SUBDIRECTORIES

IF YOU EXAMINE a listing of the subdirectories in your root directory, it will probably look something like this:

```
dr-xr-xr-x.   2 root root  4096 Feb  9 19:45 bin
dr-xr-xr-x.   6 root root  1024 Feb  9 18:39 boot
drwxr-xr-x.  18 root root  3320 Feb 11 09:26 dev
drwxr-xr-x. 117 root root 12288 Feb 11 09:26 etc
drwxr-xr-x.   3 root root  4096 Nov 12 18:31 home
dr-xr-xr-x.  20 root root 12288 Feb  9 19:45 lib
drwx------.   2 root root 16384 Nov  2 21:26 lost+found
drwxr-xr-x.   2 root root    40 Feb 11 09:26 media
drwxr-xr-x.   2 root root  4096 Jul 29  2011 mnt
drwxr-xr-x.   2 root root  4096 Jul 29  2011 opt
dr-xr-xr-x. 138 root root     0 Feb 11 09:26 proc
dr-xr-x---.   5 root root  4096 Nov 12 20:10 root
drwxr-xr-x.  33 root root  1060 Feb 11 09:29 run
dr-xr-xr-x.   2 root root 12288 Feb  9 19:45 sbin
drwxr-xr-x.   2 root root  4096 Jul 29  2011 srv
drwxr-xr-x.  12 root root     0 Feb 11 09:26 sys
drwxrwxrwt.  13 root root  4096 Feb 11 09:28 tmp
drwxr-xr-x.  12 root root  4096 Nov  2 21:28 usr
drwxr-xr-x.  17 root root  4096 Nov  2 21:31 var
```

Let's discuss some of the more important directories here. Since Linux distributions are free to modify Linux however they like, note that some of the purposes of the directories may change between versions of Linux.

The /bin directory stands for system binaries, and contains numerous system files for Linux. Many of the commands we'll discuss in this book are actually stored here.

The /boot directory stores the files necessary for Linux to boot.

The Linux kernel file, named vmlinuz-VERSION NUMBER, is located here.

The /dev directory contains the files that represent the various hardware devices on your system.

The /etc directory holds most of the configuration files for the various services and programs that run on Linux. If you administer a Linux system, you'll spend a great deal of time working in the /etc directory.

The /home directory contains the home folders for all of the users on the system. Every user that has logged into the system will have a home folder here. Usually, the home folders contain a standard set of folders – Desktop, Documents, Pictures, and so forth, along with any other individual folders the users might have created.

Both the /media and the /mnt folders are used to store removable media. We'll discuss this more in Chapters 20 and 21, but when you plug a removable disk into your computer, Linux mounts it as a folder within the filesystem. (This is different from DOS and Windows, which use letters to represent the different drives attached to a system.) Depending on the distribution, the mounted drives can show up either in /media or /mnt. Occasionally a distribution will use a /Volumes directory, but this is rare.

The /root subdirectory contains the home folder of the root user. Note that the root user is not the same thing as the root directory! We'll discuss the root user more in Chapters 7 and 8, but to summarize, the root user is the most powerful user on a Linux system, with the rights to run any program and delete or modify any file. The /root subdirectory, therefore, is the home folder of the root user, separate from the home folders in the /home directory.

The /sbin directory, like the /bin directory, holds system commands. However, the /sbin directory usually stores commands that only the root user can run, since these commands affect the entire system.

The /tmp directory stores temporary files, whether from applications or users. Note that it's a poor idea to store important data in /tmp, since some distributions regularly delete the contents of the /tmp directory when shutting down or logging off a user.

The /usr directory stores multi-user applications. While the utilities in /sbin tend to be restricted to root, applications that anyone can use often wind up in the /usr directory and its subdirectories. So most of the user applications installed on a Linux system (the Firefox web browser, the LibreOffice office suite, and so forth) are stored here.

Finally, the /var directory stands for "variable", and is used to store data that changes during the operation of the system. All of your computer's log files are stored here, along with other items like the print spooler, user mailboxes (if you have a mail server on your Linux system), and temporary files. On some distributions, if your Linux system is running a web server, the website files are stored here.

ABSOLUTE AND RELATIVE PATHS

A FINAL POINT TO know about the filesystem is the difference between relative and absolute paths.

A "file path" refers to the location of a file within the filesystem, specifically its position in the directory hierarchy. If the jmoeller user has a file named Cainai.doc in his Documents folder, then the absolute path to the file is /home/jmoeller/Documents/Cainai.doc. So if you wanted to delete the file, then the rm /home/jmoeller/Documents/Cainai.doc command would delete the file for you.

A relative path has to do with the current working directory. When you launch a command with a filename, the command first looks for the file in your current working directory. If you used the rm Cainai.doc command while /home/jmoeller/Documents was your working directory, the rm command would look in the working directory, find the file, and delete it. If your current working directory was the root directory, the command would not be able to find the file, and would then fail.

Remember that you can use relative file paths if the file or directory you intend to work with is in your current working directory. If not, you should use the absolute file path.

4

COMMANDS FOR WORKING WITH FILES AND DIRECTORIES

In the last chapter, we explained the underlying concepts of the Linux filesystem. In this chapter, we'll take a look at the commands that let you work with both files and directories. You will learn how to create and modify both directories and files from the command line. And you'll also learn how to delete files and remove directories.

FINDING THE WORKING DIRECTORY

As we mentioned in Chapter 2, the bash shell is usually configured to show the current working directory in the prompt. This isn't always the case, and if the prompt isn't configured to show the working directory, it is easy to become confused as to your current location. You can avoid this with the pwd command. Type this command:

pwd

The bash shell will respond with your current working directory.

CHANGING THE WORKING DIRECTORY

WE'VE TALKED a great deal about the working directory, but we haven't yet mentioned how to change the working directory. To do that, you need to use the cd command. (The "cd" stands for "change directory.") The cd command works with absolute file paths. If you are in the root directory, and you want to change to the /var/www directory, you would use this command:

cd /var/www

The working directory will change from / to /var/www.

The cd command also works with relative filenames in the working directory. In our previous example, you started out in the root directory. You could change to any one of the root directory's subdirectories without using the absolute file path. To change to the /etc directory, you would need only type this command:

cd /etc

The cd command also supports another useful shortcut. In the bash shell, two period characters (..) represent the directory above your current directory – in other words, the parent directory of the current working subdirectory. So if you used cd /etc to move from the root directory to /etc, you could use this command to move right back:

cd ..

Finally, there is another useful trick you can use with the cd command. Remember how the tilde character (~) represents the current user's home directory? No matter where you are in the file system, you can always return to your home directory with this short command:

cd ~

CREATING NEW DIRECTORIES

At some point, you will need to create new directories, whether to better organize your files, or as part of installing an application. To create a new directory, you will use the mkdir command. This example command will create a new directory named "ghosts" in the working directory:

mkdir ghosts.

By default, mkdir creates the new directory in the current working directory. To create the directory in a different location, either change the working directory to the desired location, or use mkdir with an absolute file path. If you wanted to create the "ghosts" directory in the Documents directory of user jmoeller, you would use this command:

mkdir /home/jmoeller/Documents/ghosts

REMOVING DIRECTORIES

From time to time you will need to remove an unused directory. You can do this with the rmdir command. To remove the "ghosts" directory we created above:

rmdir ghosts

Like mkdir, rmdir by default looks for a directory to remove in the current working directory. And like mkdir, you can use rmdir with absolute file paths:

rmdir /home/jmoeller/Documents/ghosts

An important caveat – rmdir only functions if the target directory is empty of any files or subdirectories. If you want to use rmdir to delete a directory, you must first delete the directory's contents. To delete a directory that contains files or additional directories, you will need to use the rm command, which we will discuss later in this chapter.

LISTING THE CONTENTS OF DIRECTORIES

THE PURPOSE of directories is to store files, but that won't do you much good if you can't see the contents of those directories. The ls command lets you view the contents of the current working directory. Simply use the ls command at the prompt without any options:

ls

The output will probably look something like this:

By default, ls lists the contents of the current working directory. You can also use ls with absolute file paths. For instance, if you wanted to view the contents of the /etc directory:

ls /etc

In its default mode, ls doesn't display very much information, only the names of the files and directories in the working directory. The names of the directories are usually colored-coded in blue, but not all distributions do this. How then can you get more information with the ls command?

You can do so by using ls with the -l option. "Options" are modifiers that alter the behavior of a command. They are usually (but not always) single letters added after a dash. To use ls with the -l option, type this command:

ls -l

The resultant output will look something like this:

```
[jmoeller@fedora16test ~]$ ls -l
total 36
drwxr-xr-x. 2 jmoeller jmoeller 4096 Nov 12 18:32 Desktop
drwxr-xr-x. 2 jmoeller jmoeller 4096 Nov 12 18:32 Documents
drwxr-xr-x. 2 jmoeller jmoeller 4096 Nov 12 18:32 Downloads
drwxr-xr-x. 2 jmoeller jmoeller 4096 Nov 12 18:32 Music
drwxr-xr-x. 2 jmoeller jmoeller 4096 Nov 12 18:32 Pictures
drwxr-xr-x. 2 jmoeller jmoeller 4096 Nov 12 18:32 Public
drwxr-xr-x. 2 jmoeller jmoeller 4096 Nov 12 18:32 Templates
-rw-rw-r--. 1 jmoeller jmoeller  181 Feb  9 19:57 test
-rw-rw-r--. 1 jmoeller jmoeller    0 Feb  9 18:53 Test Text File.txt
drwxr-xr-x. 2 jmoeller jmoeller 4096 Nov 12 18:32 Videos
```

The -l option stands for "long listing", and it shows far more information than the basic ls command. In the output, you can see the permissions and ownership of the files (we'll discuss permissions and ownership in chapters 11 through 14), along with their size in bytes, and the date and time they were last modified.

The ls -l command is even more useful when combined with the -h option:

ls -lh

The command will produce an output that looks like this:

```
drwxr-xr-x. 2 jmoeller jmoeller 4.0K Nov 12 18:32 Desktop
drwxr-xr-x. 2 jmoeller jmoeller 4.0K Nov 12 18:32 Documents
drwxr-xr-x. 2 jmoeller jmoeller 4.0K Nov 12 18:32 Downloads
drwxr-xr-x. 2 jmoeller jmoeller 4.0K Nov 12 18:32 Music
drwxr-xr-x. 2 jmoeller jmoeller 4.0K Nov 12 18:32 Pictures
drwxr-xr-x. 2 jmoeller jmoeller 4.0K Nov 12 18:32 Public
drwxr-xr-x. 2 jmoeller jmoeller 4.0K Nov 12 18:32 Templates
-rw-rw-r--. 1 jmoeller jmoeller  181 Feb  9 19:57 test
-rw-rw-r--. 1 jmoeller jmoeller    0 Feb  9 18:53 Test Text File.txt
drwxr-xr-x. 2 jmoeller jmoeller 4.0K Nov 12 18:32 Videos
```

This is identical to the ls -l output, save that the file sizes are now in kilobytes and megabytes, instead of bytes, which makes the output easier to read.

In Chapter 3, we mentioned hidden files. In Linux, hidden files are created by prefixing a period character (.) to the beginning of the file. Using ls with the -a switch, you can view hidden files:

```
[jmoeller@fedora16test ~]$ ls -a
.               .cache      .gconf              .mozilla
..              .config     .gnome2             Music
.abrt           .dbus       .gtk-bookmarks      Pictures
.bash_history   Desktop     .gvfs               Public
.bash_logout    Documents   .ICEauthority       .pulse
.bash_profile   Downloads   .imsettings.log     .pulse-cookie
.bashrc         .esd_auth   .local              .setroubleshoot
```

COPYING FILES

Now that you know how to create directories, you may want to put files into those directories. There are a number of commands you can use to move files, and we'll start with the cp command. The "cp" is short for "copy", and allows you to copy files from one location to another. When selecting the file you want to copy, you can use the relative file path if it's in the current working directory, but the destination must be an absolute file path. To copy the file Caina1.doc from the current working directory to the /backup folder, you would use this command:

cp Caina1.doc /backup

This will create a copy of Caina1.doc in the /backup folder.

Like most commands, cp assumes that you are copying files from the working directory, and looks there first. However, you can also specify the source file with an absolute file path:

cp /home/jmoeller/Documents/Caina1.doc /backup

What happens if there's already a file named Caina1.doc in the target directory? If there's already a file there, the cp command will pause for your confirmation before overwriting the file. Hit the Y key when prompted, and cp while overwrite the file with the new data. Be careful when copying files, since you might accidentally overwrite valuable data.

The cp command also works with wildcards, which we will discuss at the end of this chapter.

MOVING AND RENAMING FILES

To move a file, you use the mv command. Like the cp command, the mv command lets you use a relative file path as the source file, but you will need an absolute file path for the destination. Let's say you wanted to mv the Caina1.doc file from the current working directory to the /backup folder. You would use this command:

mv Caina1.doc /backup

The command will then move the Caina1.doc file from the current working directory to /backup.

Like cp, you can also use mv with an absolute file path for the source file:

mv /home/jmoeller/Documents/Caina1.doc /backup

The mv command also has an interesting secondary function – you can use it to rename files from the command line. This works because you can specify that the moved file have a different name at the end of the operation. In essence, you've moving a file to the same directory but with a different name. To rename the Caina1.doc file to Caina2.doc, you would use this command:

mv Caina1.doc Caina2.doc

Assuming that Caina1.doc is in the current working directory, the file will be renamed.

The mv command also works with wildcards, which we will discuss at the end of this chapter.

DELETING FILES AND DIRECTORIES

Copying and moving files is useful, but sometimes you simply need to delete them. You can do this with the rm command. Like the cp and mv commands, rm works with both absolute and relative file paths. To delete a file named Caina1.doc in the current working directory, you would use this command:

rm Caina1.doc

The command will then delete the Caina1.doc file. Likewise, you can use rm with an absolute file path:

rm /home/jmoeller/Documents/Caina1.doc

Note that you will need the appropriate permissions to a file in order to delete it. (We'll discuss permissions more in Chapters 13 and 14.) If you do not have the required permissions to a file, you will not be able to delete it.

As we mentioned in the section on the rmdir command, you can use the rm command to delete directories that still contain files and subdirectories. You can do this by modifying rm with the -R option. If you wanted to delete a directory called /home/jmoeller/storage that still held files, you would use this command:

rm -R /home/jmoeller/storage

The rm command will then delete /home/jmoeller/storage and any files it still contains. The command will also delete any subdirectories /home/jmoeller/storage contains, along with any child directories within those subdirectories. Needless to say, you should be careful when using rm with the -R option! You can wipe out entire branches of the filesystem and lose important data if you're not careful.

(Note that the -R option also works with the cp command, so you can likewise copy large chunks of the filesystem as once.)

The rm command also works with wildcards, which we'll discuss at the end of this chapter.

ARCHIVING AND COMPRESSING FILES

MANAGING a large number of files and directories can prove tricky, especially if they're spread across several directories. It's much easy to take those files and create a single archive – a single large file created out of numerous small files, which you can later unpack back into the individual files. It's also useful to compress files – compressed files are packaged in such a way that they take up less space on the hard drive. Modern hard drives are enormous, so compressing files may not seem as critical as it once did. However, many files you download over the Internet are compressed, so you'll need to know how to uncompress them. And when you upload files to the Internet or attach them to an email message, you may need to compress them to save space.

You can use the tar command to create archives from the command line. Archives created with the tar utility generally have a .tar extension, and are often nicknamed "tarballs." To create an archive with tar, you'll need to use both the -c option, to create an archive, and the -f option, to specify the name of the archive. The following command will create an archive named storage.tar of your ~/Documents folder in the current working directory:

tar -cf storage.tar ~/Documents

(It's considered good form to create tar archives using a directory that contains files, rather than a number of individual files. That way the archive unpacks into a single directory, rather than many individual files.)

This will create the storage.tar file in the current directory. When working with tar, always remember to specify the .tar extension for the archives you create, since tar will not do so automatically. (Note that tar does not alter the original files in anyway – it merely creates an archived copy of them.)

However, this new archive file is not compressed, and takes up the exact same amount of space as your ~/Documents folder. To compress it, you need to use tar with the -z option, which compresses the archive with gzip compression. The following command will create a compressed archive:

tar -czf storage.tar.gz ~/Documents

You might also want to add the -v option as well, for "verbose" output, which means tar tells you exactly what it is doing and lets you know if something goes wrong:

tar -czvf storage.tar.gz ~/Documents

Now that you have files neatly packed up into an archive, you might want to unpack them. To unpack files from a tar archive, you use the -x option to eXtract them, rather than the -c option to compress:

tar -xvf storage.tar

If a tar archive has been compressed with gzip compress, make sure to add the -z option to the command when you wish to uncompress it:

tar -xzvf storage.tar.gz

The tar command is useful in a variety of situations, but it is rather complex. A simpler way of archiving and compressing your files is with the zip and unzip commands. To cite our previous examples, if you wanted to create a zip file of your ~/Documents directory, you would use this command:

zip -r backup.zip ~/Documents

This will create the backup.zip file in the current working directory.

To uncompress the zipped archive, you will use the unzip command:

unzip backup.zip

Note that the files are uncompressed in the working directory. Alternatively, if you use unzip with a absolute file path, the files are uncompressed in the same directory as the source zip file.

WILDCARDS

WE'VE DISCUSSED COPYING FILES, moving files, renaming files, and deleting files, but now we'll describe a feature of the shell that allows you to make your file manipulation commands vastly more powerful. Usually, when you employ the cp or the rm commands, they only affect one file at a time. But if you use wildcards in conjunction with the file manipulation commands, you can affect hundreds or even thousands of files at the same time.

What are wildcards? "Wildcards" are special characters that represent multiple other characters, or even all other possible characters. The first wild card character, the question mark (?), represents one potential character. For example, let's say you had seven files in your current working directory named Caina1.doc, Caina2.doc, and so forth. If you wanted to delete them, you could use the rm command to remove them one at a time. Or, you could use rm in conjunction with a wildcard:

rm Caina?.doc

This will delete every file in the working directory that begins with "Caina", has an extension of .doc, and has a single character after "Caina." You can use this character with cp, mv, ls, and other commands as well – you could copy the Caina?.doc files, or move them, or use it with ls to only view those files.

The second wild card character, the asterisk (*), is even more powerful. The question mark wildcard character only represents a single character. The asterisk represents any number of characters in any combination. For example, if you wanted to delete every single .doc file in your current working directory, you could delete them one by one. Or, you could use the asterisk wildcard character with rm:

rm *.doc

This will delete every single file with the extension of .doc in the current working directory. Needless to say, you should exercise caution when using wildcards with the rm command! You can also use the asterisk with the other command – with ls, for instance, to only view the .doc files in your current directory, or with the cp command to copy only the .doc files.

Wildcards allow you to make the file manipulation commands vastly more powerful. You can also use them with the file permission and ownership commands, which we will discuss in Chapters 11 through 14.

5

CONCEPT: USERS AND USER ACCOUNTS

To use your Linux system, you have to enter a username and a password. This is your user account, and user accounts are an important part of the Linux command line. In this chapter, we'll examine the underlying concepts of user accounts on Linux.

WHAT IS A USER ACCOUNT?

To put it simply, a user account is the method an operating system uses to determine which users have access to which files and programs. In Linux, files and programs are configured to be secure – users cannot view other users' home directories, for instance, and only the root user can delete and alter important system files and programs. How does the operating system know to let the current user of the computer have access to the appropriate files and programs? User accounts are the method used to grant access. A user presents his username, and then a password to confirm that he is indeed the user in question. (Technically, this is called single-factor authentication, since you only need a single piece of information, the password, to prove that you are a specific user. A complex kind of authentication, where you need a password and a PIN number to log in, is called two-factor authentication.)

To properly secure a Linux system, you need to have the user accounts configured correctly.

TWO KINDS OF USERS

On most kinds of Linux systems, you will see two kinds of users – standard users, and admin users.

Standard users are the normal users on the system. They have full access to their individual home folders, and can run the programs designated for normal users. They cannot view the home folders of other users, and cannot alter or change system settings or installed programs. The employees of an accounting department, for instance, would usually be standard users – they have access to their files and the programs they need to do their work, but cannot change the system.

Admin users are just like normal users, but with one exception. They have the right to either use or borrow the powers of the root user (using a command called sudo, which we will describe in Chapter 8) to make changes to the system's settings and programs. The IT professionals supporting the accounting department, to continue our example, would generally be admin users.

FILES CONTROLLING USER ACCOUNTS

IN LINUX, user accounts are stored in one file, and passwords in another.

The user accounts in your system are defined in the /etc/passwd file. Despite the name, the passwords are not actually stored in the /etc/passwd file. Older versions of Linux used to store the passwords there, but that presented a security risk, so the passwords were encrypted and moved to a different file. There is still quite a bit of useful information in the file. If you view it from the command line (we'll show you how to do that in Chapter 16), the entries will look something like this:

```
jmoeller:x:1000:1000:Jonathan Moeller:/home/jmoeller:/bin/bash
```

Colons separate the specific fields in the entry.

The first piece of information is the username (in this case, jmoeller). The second field has only an x. In older versions of Linux, the password would go here, but passwords are no longer stored in /etc/passwd. The third field holds the user account's numerical user ID number (commonly shortened to UID). In this case, the UID is 1000. Most Linux distributions give the first new account on the system a UID of 1000, and then count up from there.

The fourth field contains the account's group ID number (commonly shorted to GID). Generally, each new account receives its own group (called a primary group) for reasons of security. We'll discuss groups more in Chapters 9 and 10.

The fifth field holds the user's full name (in this case, Jonathan Moeller), the sixth holds the location of the user's home directory, and the seventh stores the location of the user's default shell. Most Linux distributions use the bash shell as the default shell, so this final entry will almost always be /bin/bash.

You might be tempted to add accounts to your system by manually editing /etc/passwd. Do not do this! Misconfiguring /etc/passwd can cause serious problems for your system. Always use the appropriate command line utilities to add user accounts to your system, which we shall discuss in the next chapter.

If user accounts are stored in /etc/passwd, user passwords are actually stored in the /etc/shadow file. You need root permissions to view the /etc/shadow file, but if you view it, each of the entries will look something like this:

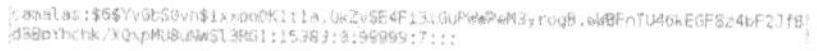

Like with the /etc/passwd file, each of the entries in /etc/shadow are divided by colons into separate fields. The first field contains the username (in this case, camalas).

The second field contains the password in encrypted hash form, hence the long string of numbers and letters. If there's only an asterisk in the second field, that means the account in question is not actually allowed to log into the system. You'll usually see this for system user accounts (accounts that system services use to access the system) or regular user accounts that have been locked.

The third field contains the number of days since January 1st, 1970, since the password was last changed. In this example, the number of days is 15,383, which means it has been about 42 years since the password was last changed. If you do the math, the password in this example was last set on February 12th, 2012. This is a cumbersome way to go about it, but remember that Linux was derived from UNIX, and Linux has carried over UNIX's password system.

The fourth field is the minimum number of days before a password can be changed. This is to enhance password security policies – if users are forced to change their passwords, this prevents them from changing their passwords right back. If there's a 0 in the field, as in this example, that functionality is disabled.

The fifth field sets the maximum number of days before a password expires and must be changed. Setting it to 99,999, as in this example, disables this feature. (Since 99,999 days is about 273 years, the issue is moot!) So the password of the camalas user account will never expire.

The sixth field specifies the number of days before the password expires that a warning will be issued. The default setting is seven days, so if a password is set to expire, the user will receive a warning while logging in for seven days.

The seventh field specifies the number of days after January 1st, 1970, to wait until locking the account after the password has expired. So if a password expires, this field specifies the number of days to wait until locking out the user. If you see a value of -1 in the field, as in this example, that functionality has been disabled.

The eighth and final field specifies the number of days since January 1st, 1970, to wait until disabling the user account. If the field is blank, as in this example, that means the account is set to never be disabled.

As with /etc/passwd, you might be tempted to modify the passwords by editing the file. Again, do not do this! Both /etc/passwd and /etc/shadow need to remain in sync with each other, and incorrectly editing either file might damage your Linux system. It is better to use the command line utilities for changing passwords, which we will describe in the next chapter.

THE ROOT USER

THE ROOT USER on a Linux system has access to any file and can delete any file or directory. Needless to say, logging in as the root user should be done with caution, as you can accidentally damage or disable your Linux system. We will discuss how to use the root account safely in Chapters 7 and 8.

6

COMMANDS FOR WORKING WITH USER ACCOUNTS

Now that you understand the principles behind user accounts, we'll explain how to use command line utilities to control them. In this chapter, you'll learn how to create, modify, and delete user accounts. You'll also learn how to reset passwords and lock accounts.

CREATING USER ACCOUNTS

As we mentioned in the previous chapter, adding user accounts by manually editing the /etc/passwd and /etc/shadow files is a bad idea. Fortunately, you can use the useradd command to add accounts to your system from the command line. Let us say you wanted to add an account for a woman named Caina Amalas to your system, and have decided to assign her a username of camalas.

(Shameless authorial plug: Caina Amalas is the name of the protagonist from my THE GHOSTS series of fantasy novels and short stories.)

To begin, it's best to choose a naming convention for your user accounts. A "naming convention" sounds fancy, but it just means that you stick to a consistent way of naming the accounts. Usually, something like the first initial and the last name is best – like "camalas" for Caina Amalas. Letting users pick their own usernames in a business setting is a bad idea, since users will very often choose inappropriate or even obscene usernames. (To cite an actual example I have encountered, letting a user in the sales department give herself a username of partygrrl19 does not reflect well on the business.)

To add the camalas account to your system, you would use this command:

useradd camalas

(Note that you need to root-level permissions to add user accounts to your system, and we will discuss using root powers more in Chapters 6 and 7.)

This command will add the camalas user account. However, useradd without any options is not terribly useful. The command creates the camalas user account, but it doesn't assign a display name for it, it doesn't create a home directory for camalas, and it doesn't set other important options – it neglects to assign a default shell for the account, for instance.

We'll have to add some options to make useradd more effective. First, use the -m option create a home directory for the camalas account:

useradd -m camalas

The -m option will add a home directory for camalas in the /home directory. Next, you'll want to use the -s option to set a default shell from the account:

useradd -m -s "/bin/bash" camalas

This sets the bash shell as the default shell for the camalas account. There's one final switch to add to the useradd account – the -c option to set the account's display name as "Caina Amalas."

useradd -c "Caina Amalas" -m -s "/bin/bash" camalas

With these three options, the useradd command creates a new account called camalas, sets the account's display name as Caina

Amalas, creates a home directory for the account in /home, and sets bash as the account's default shell. Finally, you need only set the password for the account with the passwd command:

passwd camalas

(That isn't a typo – the proper command is in fact "passwd", not "password.")

The camalas account is now ready for use. Note that to change another user's password, you need to run passwd as root or with root-level permissions.

MODIFYING USER ACCOUNTS

Once you've created a user account, you may need to modify it. In this section we'll first describe how to change user account passwords, and how to modify, lock, and otherwise alter existing user accounts.

Changing the password for your own user account is simple enough. Simply type the passwd command:

passwd

You'll receive a prompt asking for your current password. After you enter it, the command will then prompt you for your new password, and then prompt you again for confirmation. Your new password will be set.

To reset the password of a different user, you still use the passwd command. Note that in order to change another user's password, you need to either be root or running with root-level permissions. As we mentioned in the previous section, to change the password for the camalas account, you would use this command:

passwd camalas

You will not need to know camalas's current password. Simply enter the new password twice when prompted, and the command will change the password for the camalas account.

From time to time you may find it necessary to lock out an account, preventing the user from logging in. This very often happens is a user is misbehaving or abusing his or her computer privileges. On a less negative note, this can happen if you have a contract employee whose contract expires, but the company plans to rehire in a few months. To do so, you will use the usermod command, which is like the Swiss Army knife of user account management. To lock out the camalas user account, use usermod with the -L switch.

usermod -L camalas

(Once again, to use any of the usermod command's functions, you'll need root-permissions. Note also that the -L switch must be uppercase – remember that the Linux command line is case sensitive.)

This will lock the account, leaving camalas unable to log in. To unlock the account, use usermod with the -U switch. (Note that the -U must be capitalized.)

usermod -U camalas

The camalas account will be unlocked.

The most common other change you'll need to make to individual user accounts (other than changing their group membership, which we will discuss in Chapters 9 and 10) is to change the name of a user account. The most common reason for this is when a woman gets married and changes her name. To continue our previous example, let's say Caina Amalas marries a man with the last name of Maraeus and changes her name to Caina Maraeus. She then wants to change her username to cmaraeus and her display name to Caina Maraeus.

You can accomplish both these tasks with the usermod command.

To change the username of the account from camalas to cmaraeus, use usermod with the following options:

usermod -l cmaraeus camalas

This will change the account's username from camalas to cmaraeus. It might seem strange to have cmaraeus before camalas, but the command will not work otherwise.

To change the display name of the camalas account, use the

usermod command with the -c switch:

usermod -c "Caina Maraeus" camalas

(Note that if you've already changed the account name to cmaraeus, make sure to use that instead.)

The account's display name will change to Caina Maraeus instead of Caina Amalas.

DELETING USER ACCOUNTS

Finally, from time to time you might find it necessary to delete user accounts. Perhaps an employee leaves your organization permanently, or perhaps an employee no longer needs access to a particular system. Most of the time, it's a better idea to simply lock a user account than to delete it entirely. Just because an employee leaves an organization doesn't meant that he or she will not return in the future, after all. And if the employee does return, it's a lot easier to unlock a locked user account than to recreate it from scratch.

But if you must delete a user account, you can do so with the userdel command. If you wanted to delete the camalas account, the command would look like this:

userdel camalas

(Note that you must be logged in as root or have root-level powers to delete user accounts.)

This will delete the camalas account, but it will not delete the account's home directory in /home. If you want to preserve the user's files, this is the best course to take. But if you're absolutely sure that you need neither the camalas account nor the user's files ever again, you can delete both the account and the home folder with this command:

userdel -r camalas

This command will get rid of both the camalas account and the /home/camalas directory (along with all its subdirectories).

7

CONCEPT: THE ROOT USER

We've already mentioned the root user several times. In this chapter, we'll take a closer look at the root user and its capabilities. In the following chapter we'll show you how to use the root account safely, without damaging your system.

WHAT IS THE ROOT ACCOUNT?

THE ROOT ACCOUNT, also known as the superuser account, is the most powerful account in a Linux system. The root account has permission to access, modify, and delete any file. The root user can also install and uninstall programs, modify system-wide settings, and change the ownership of files (we'll discuss ownership more in Chapters 11 and 12).

The name comes from the early days of UNIX – the root user was the only user with the ability to change the root directory, and so the term carried over to the superuser account.

The root user's entry in /etc/passwd will usually look something like this:

```
root:x:0:0:root:/root:/bin/bash
```

The root account will always have a UID of 0, and a group ID of 0. The home directly for root is usually /root.

USING THE ROOT ACCOUNT SAFELY

IF YOU ARE the administrator of a Linux system, it might be temping to use the root account as your day-to-day account. After all, if you are performing system tasks, sooner or later you will need to use root, so why not use it as your main account? Do not give in to this temptation. The superuser account has control over every aspect of the system, and even a simple typographical error in a command issued by root can be devastating. It is a good idea to make yourself a standard user account, and use root only when necessary.

Of course, logging out and logging back in every time you need to perform an administrative task will quickly prove cumbersome. To counter that, a command called su lets you log in as root from within your own account. Once you are finished in your root session, you need only type exit to return to your own user account.

What happens if you work on a system where multiple people need access to the user account? Sharing the root password with multiple people is a security risk (since those people might, in turn, share the password with others). It also makes auditing and tracking system changes more difficult. If the root user makes a change to the system, it's hard to figure out who is responsible if five different people know the root password.

To get around this problem, Linux offers the use of the sudo command. The sudo command lets you temporarily "borrow" the powers of the root user to perform a task. Only users belonging to a specific group (usually the admin group) can use the sudo command, and their actions are recorded in a log file.

In fact, with the sudo command, you can even keep the root account disabled and still use root-level powers. Keeping the root account activated is something of a security risk, since if an attacker guesses your root password, he has complete control over your entire system. Some distributions (like Ubuntu) disable the root account by default, and recommended that administrators use the sudo command to perform administrative tasks.

In the next chapter, we'll take a closer look at using the su and sudo commands.

8

COMMANDS FOR USING THE ROOT ACCOUNT

In this chapter, we will explain how to use root's powers safely via the su and the sudo commands.

THE SU COMMAND

THE "SU" command stands for "switch user." In its default state, it prompts you to log into the root account. However, you can actually use su to switch to any account on the system for which you know the password.

First, though, we'll discuss root. To switch to the root account, simply type su without any switches:

su

Enter the root password, and su will switch you to the root account without terminating your session under your own account. Once you've finished using the root account, type the exit command. You'll log out of root, and return to your own session.

You can also use su to log in as any other user in the system (assuming you know their password – if you don't, you must first use the passwd command to change their password). For example, if you want to log in using the camalas account from the previous chapter, use this command:

su camalas

Enter the password for camalas, and you'll log in using that account. Once you are finished, you can close camalas's session by using the exit command.

THE SUDO COMMAND

THE SUDO COMMAND offers a more sophisticated way to deal with root powers. The "sudo" command stands for "switch user do" - though because it's so often used for administrative tasks, it's often called the "superuser do" command. With su, you have to log in completely as root. The sudo command only "borrows" the powers of root for a brief time to complete a specific task, and then completes. This means you can perform administrative tasks from your regular user account, so long as your account has permission to use sudo.

For example, let's say you wanted to use the usermod command to lock the camalas account from the session of a standard account. Type this command:

usermod -L camalas

You will receive an access denied error message.

However, if you preface the command with sudo, you will get better results:

sudo usermod -L camalas

Before executing the command, sudo will ask for your password. Enter your password, and sudo will check to see if your password is on the list of accounts allowed to use sudo. If it is, sudo will execute the command as root.

We mentioned the "admin" group earlier in this book. In most Linux distributions, the admin group is the one that has the right to use sudo to employ root powers. The list of accepted sudo users is defined in the /etc/sudoers file. (Appropriately enough, you need root rights to view the /etc/sudoers file.) In most distributions, the list of sudoers consists of the admin group and nothing else.

The use of sudo is also recorded in a log file. In most distributions, sudo use is record in the /var/log/auth.log file. Scanning this file will let you see who is using sudo and for what tasks.

This is a book about the command line, but sometimes you might need to run a graphical application as root. You can do this using the gksudo command – use the gksudo command in a terminal emulator application, and the application you've launched with gksudo will appear in a window in the GUI.

9

CONCEPT: GROUPS

The last four chapters should have given you a solid grasp of working with user accounts from the Linux command line. But user accounts are only aspect of managing user accounts. When you sort your various user accounts into groups, you can far more easily assign the appropriate permissions. In this chapter, we will explain the underlying principles of Linux groups.

WHAT ARE GROUPS?

In Linux (and many other operating systems), a "group" generally refers to a group of user accounts. The basic purpose of groups is to simplify the management of permissions to files and folders. For example, let's say you have a file named BUDGET.DOC that every member of the accounting department needs to access. You could assign permissions to every single member of the accounting department, but that could quickly become cumbersome, and for larger organizations (where the accounting department might have a hundred people) such a manual method would quickly become unmanageable.

Groups offer a way to simplify this mess. You need only create a group called "accounting", and assign all members of the department to it. Then if you assign permissions to BUDGET.DOC to the accounting group, every single user in the accounting group gets permission to access the file, no matter how many users are in the group. So if a hundred different people all need access to a file, it's much easier to assign the permission to their group just once rather than a hundred different times to a hundred different users.

PRIMARY GROUPS

As you add accounts to your system, you will quickly realize that every time you create an account, a group with the same name is also created. Furthermore, the new user account is always added to the new group of the same name, and is usually the only member in the group. These are generally called "primary groups", and are created because every Linux user must be a member of at least one group, but they are used for reasons of security. We will discuss ownership and permissions in more detail in the next four chapters, but every file and directory in Linux is assigned an "owning" user and and an owning group. ("Ownership" is Linux simply means that the owner of the file or the directory has full control over it.) Often you will have files that you want only a single user to control and modify. If the file's ownership is set to the user's account and the "users" group containing every group on the system, security quickly becomes a problem. But by assigning ownership to the user's account and the user's primary group, only that user (and root, of course) will have access to that file or directory.

USERS AND ADMIN GROUPS

MOST LINUX DISTRIBUTIONS will have two default groups for user accounts – the users group, and the admin group. The users group will contain all the standard users accounts on the system. If you want every user on the system to have access to a file or a directory, you would assign the group ownership to the users group. (We'll discuss how to do that in Chapter 12.)

We've already mentioned the admin group in connection with the sudo command and the superuser account. On Linux distributions that use sudo, the admin group is the group that gets the right to use sudo. Usually, the admin group is added to the /etc/sudoers file. (This is an example of using groups to simply administration – rather than manually add users to the /etc/sudoers file, you need only add them to the admin group.)

Most Linux systems will have a variety of other groups related to system functions. There's a "cdrom" group, for example, that specifies who has the right to use the optical drive on the system. There are also a number of groups related to system services and background programs. Generally, if you're new to the Linux command line, it's best to avoid tampering with these groups unless you know what you'rc doing.

GROUP ID AND CONFIGURATION FILES

USER ACCOUNTS GET a user ID (UID) number, and groups, likewise, receive a GID – a group ID number. Like UIDs, GIDs follow some standard conventions.

The user group for root usually receives a GID of 0. The standard users group will receive a GID of 100. The GIDs for primary groups, like the UIDs for user accounts, starts at 1000 and works its way up – the first primary group on the system will have a GID of 1000, the second a GID of 1001, and so forth. GIDs of below 1000 are reserved for various system groups.

Just as user accounts are defined in the /etc/passwd file, groups are defined in the /etc/group file. A typical entry in the /etc/group file will look like this:

```
adm:x:4:jmoeller
```

Like /etc/passwd, the individual fields in /etc/group are separated by colons. The first field denotes the name of the group. The second field stores the group password – similar to /etc/passwd, group passwords are actually stored in encrypted form in the /etc/gshadow file. The third field shows the GUID. The fourth and final field displays the members of the group. Depending on how many users are assigned to the group, this field can become rather long.

As with /etc/passwd, editing the /etc/group file by hand is a bad idea and can mess up your system. Instead, it is better to use command line utilities to add groups, add users to groups, remove users from groups, and delete groups. We'll show you how to do just that in the next chapter.

10

GROUP MANAGEMENT COMMANDS

After the last chapter, you should understand the underlying mechanics of Linux groups. In this chapter, we'll show you the commands to create groups and add users to groups. We'll also explain how to remove users from groups and delete unnecessary groups. Note that all group-related commands need to be run with root powers, whether directly logged in as root or by using sudo.

You'll find that the commands for working with groups are fairly simple. While groups can simplify complex permissions management, managing groups themselves is a straightforward task.

CREATING A NEW GROUP

To create a new group, you will use the groupadd command. If you wanted to create a new group called "ghosts", the command would look like this:

groupadd ghosts

And that's it! The command will add the ghosts group to your system. There are a number of different options available for groupadd. If you wished, you could use the -g switch to specify a GID. For the most part, the default options will work fine.

ADDING A USER TO A GROUP

NOW THAT YOU'VE created a new group, you probably want to add one or more accounts to it. You can do this using the Swiss Army knife of user account commands, the usermod command. With the -G switch, usermod will add the user account of your choice to the specified group. If you wanted to add the camalas user account to the "ghosts" group we created in the previous section, you will use this command:

usermod -G ghosts camalas

This will add the camalas account to the ghosts group.

REMOVING A USER FROM A GROUP

REMOVING a user from a group is likewise a easy task. To remove a user from a group, you will use the gpasswd command. The following command would remove the camalas account from the ghosts group:

gpasswd -d camalas ghosts

Once the command finishes, camalas will not longer be a member of the ghosts group.

DELETING A GROUP

To delete a group, you will use the groupdel command. For example, to delete the ghosts group we created in the earlier section, use this command:

groupdel ghosts

You might recall that the rmdir command requires that a directory be empty before you delete it. The groupdel command does not work the same way – you can delete a group that still contains user accounts. It is important to note, however, that deleting a group does not delete the user accounts contained within a group. (This is a useful feature, since accidentally deleting the wrong group on a large system could inadvertently wipe out hundreds of user accounts.) So you can safely delete a group that still has members without worrying about deleting anyone's user accounts.

There is one exception to this rule – the primary group of an individual user account. Remember that Linux requires a user account to be a member of at least one group. So to delete a primary group, you must first delete its associated user account.

11

CONCEPT: OWNERSHIP

In the previous chapters, we've touched a little on the concept of file and directory ownership. In this chapter, we'll explain the ideas behind ownership, and explain how they apply to file permissions. Ownership might seem complex, but combined with file permissions, it makes for an elegant way to manage file security on a Linux system.

WHAT IS OWNERSHIP?

WHAT IS file and folder ownership? We are not talking about the person who owns the computer, in the sense of the person or organization that paid for the computer hardware. Instead, when we talk about who owns the file in Linux, we are describing the user that has full control over the specific file. Basically, the user "owns" the file, and has the right to change its permissions as he or she pleases. Users own the files in their home directories. System files and configuration files are generally owned by the root user, which is why you need root powers to change many system settings. Sometimes ownership will be assigned to the user accounts of various system services, if they need to regularly access the files or directories in question.

Ownership comes into play with file permissions. Basically, every file or folder on a Linux system can be accessed by three different divisions of users – the owning user, the group associated with the owning user, and all other users on the system. Access permissions to the file are assigned on that basis – you can give the owning user full control over a file, for example, but only give read access to the associated group, and deny access to all other accounts on the system.

Understanding ownership is key to assigning correct permissions to a file. Because each file has three sets of permissions – the permissions assigned to the owning user, the permissions assigned to the associated group, and the permissions assigned to all other accounts on the system – you can quickly change a file's permissions by changing its owner or its owning group.

We will describe permissions in greater detail in Chapters 13 and 14. In the next chapter, we will describe the commands to change and modify ownership of files and folders.

12

COMMANDS FOR CHANGING OWNERSHIP OF FILES AND DIRECTORIES

In the previous chapter, we discussed the principles and concepts behind the ownership of files and directories. In this chapter we'll show you how to manage ownership from the command line. You'll learn how to determine the ownership of files, and how to change both the owning user and the associated group.

DETERMINE THE OWNER OF FILES

THE FIRST STEP to working with the ownership of files is to determine who currently owns those files. You can do this with the ls command. Recall that the ls command without any options does not display very much information, but that if you use the ls command with the -l option, much more information is displayed:

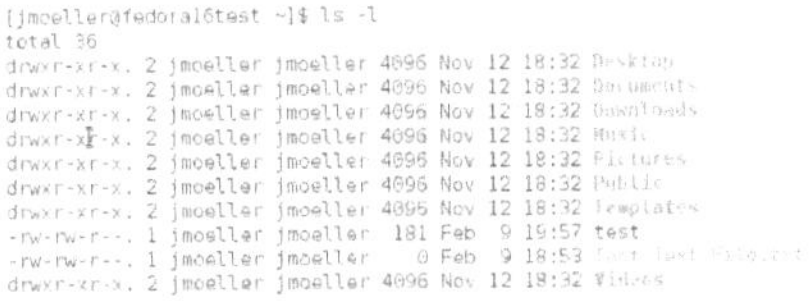

```
[jmoeller@fedora16test ~]$ ls -l
total 36
drwxr-xr-x. 2 jmoeller jmoeller 4096 Nov 12 18:32 Desktop
drwxr-xr-x. 2 jmoeller jmoeller 4096 Nov 12 18:32 Documents
drwxr-xr-x. 2 jmoeller jmoeller 4096 Nov 12 18:32 Downloads
drwxr-xr-x. 2 jmoeller jmoeller 4096 Nov 12 18:32 Music
drwxr-xr-x. 2 jmoeller jmoeller 4096 Nov 12 18:32 Pictures
drwxr-xr-x. 2 jmoeller jmoeller 4096 Nov 12 18:32 Public
drwxr-xr-x. 2 jmoeller jmoeller 4096 Nov 12 18:32 Templates
-rw-rw-r--. 1 jmoeller jmoeller  181 Feb  9 19:57 test
-rw-rw-r--. 1 jmoeller jmoeller    0 Feb  9 18:53
drwxr-xr-x. 2 jmoeller jmoeller 4096 Nov 12 18:32 Videos
```

SPECIFICALLY, the ls -l command displays the permissions, the owning user, the associated group, the file's size, the time it was last altered, and the file's name. The owning user is listed first (in this case, "jmoeller") and the associated group is listed next (in this example, the primary group for the jmoeller user account).

So long as you have the read permission for a file or a directory, you can use the ls -l command to view the owners. If you do not have any permissions to a file or a directory, you will receive an access denied error message if you try to use ls -l.

CHANGING OWNERS

SOMETIMES YOU MAY NEED to take ownership of files, or to assign them to different owners. To use our previous example, if an employee leaves the company, you will need to assign ownership of his files to a different user (and probably move those files to a new home folder, as well). Or an employee might transfer to a different department within the organization, and you will need to change the ownership of the files related to his old job.

To change file ownership, you use the chown command – the "chown" stands for change owner. Note that to change the owner of a file or a directory, you will need to log in as the root user or to posses root-level powers through the sudo command. If, for instance, you wanted to assign the ownership of a file named test.txt in the current working directory to the camalas account, you would use this command:

chown camalas test.txt

Like many other commands, chown works with both relative and absolute file paths. If the file you wish to modify is in the current working directory, you can use a relative path. If you wanted to use chown with an absolute file path, you could also do so:

chown camalas /home/camalas/test.txt

You will notice that chown only affects one file or directory at once. How do you change the ownership of a large number of files at once? Like the rm or cp commands, all the commands for managing permissions and ownership work with both the recursive -R option and with wildcards. For instance, to assign ownership of every .doc file in the /test directory to camalas:

chown camalas /test/*.doc

More often than not, you'll want to assign ownership of entire directories and their subdirectories at once. To assign ownership of the /test directory, and all its files and subdirectories, to camalas:

sudo chown -R camalas /test

Using chown, you can quickly manage ownership on your Linux system.

CHANGING GROUP OWNERSHIP

As we mentioned, every file and directory is assigned an owning group, and permissions are assigned to that group. You might find yourself having to change the owning group more often than the owning user. You'll recall from the chapter on groups that it is a lot easier to join user accounts to a group and then assign permissions based on the group, rather than individually giving permissions to each user account.

To change the group ownership of a file, use the chgrp command.

For example, if you wanted to change the group ownership of the test.txt file to the ghosts group:

chgrp ghosts test.txt

Like chown, chgrp works with either absolute or relative file paths. If the file is in the current working directory, chgrp can find it, otherwise you will need to use an absolute file path:

chgrp ghosts /home/camalas/test.txt

And like chown, chgrp also works with wildcards and the recursive -R switch. To change the group ownership of the /test directory and all its files and subdirectories, use this command:

sudo chgrp -R ghosts /test

Using chown and chgrp lets you easily change the user and group ownership of files and directories from the Linux command line. However, once you've set the ownership, you will need to assign permissions to the owner and to the group. We'll explain those concepts more in the next chapter.

13

CONCEPT: FILE AND DIRECTORY PERMISSIONS

We've mentioned the concept of permissions frequently in the last few chapters. In the next two chapters, we'll first explain the concept of permissions in the Linux filesystem. You'll learn about the three major permissions every file and directory can be assigned – read, write, and execute. You'll also learn about the numeric codes for those permissions, and the commands necessary to modify permissions.

WHAT ARE PERMISSIONS?

"Permissions", also known as "access control", are rules that define which users get to access which files, and what those users can do if they have access to those files. Permissions are absolutely necessary to secure your Linux system, especially if you have multiple users logging into the same machine. In conjunction with user and group ownership, properly assigned permissions prevent regular users from tampering with important system files and directories, and block regular users from viewing each other's documents.

HOW ARE PERMISSIONS ASSIGNED?

For purposes of assigning permissions, Linux users are divided into three separate classes – the owning user, the owning group, and all other accounts on the system. That means you can set different levels of permissions for the three different classes – you could allow the owning user to modify the file, you could allow the accounts in the owning group to only read the file without having the right to modify it, and you could block all other accounts from viewing it entirely. This is why it's important to have an understanding of ownership and groups – it does no good to assign permissions if you don't know to whom you are assigning those permissions!

In the next section we will examine the three permissions you can assign to a file.

THE THREE PERMISSIONS

The are three permissions that can be assigned to each of the classes of users – the Read permission, the Write permission, and the Execute permission.

The Read permission allows you to access the file, but not to delete it or alter it. For example, you could open a document with the Read permission, but you could not edit it, and you could not delete it. If the Read permission is assigned to a directory, you can list the files within the directory (but unless you also have the Read permission to the files themselves, you won't be able to read them). Note that having the Read permission for a directory allows you to view the file names, but no other information about them.

The Write permission allows you alter or delete a file. To return to our example document, if you have the Write permission, you could delete it or make changes to it. If you have the Write permission to a directory, you can create files within it. (Ironically, if you don't have the Read permission to a directory while you have the Write permission, you can create files within the directory, but cannot read them).

The Execute permission allows you to run a file as a program. Any application or program files need to have this permission, otherwise they won't work. If you have the Execute permission for a directory, you can cross it to view the contents of its subdirectories.

These permissions are combined to make up a file or directory's total permissions. Remember that each file or directory gets three sets of permissions – the permissions for the owning user, the permissions for the owning group, and the permissions for all other accounts on the system.

In the next chapter, we'll show you how to both view and change the permissions for files and folders.

14

COMMANDS FOR CHANGING FILE AND DIRECTORY PERMISSIONS

The previous chapter explained the concept of Linux file permissions. In this chapter you will learn how to use the command line to view and modify file permissions. We'll show you how to find a file's permissions, and how to change them.

DETERMINE A FILE'S PERMISSIONS

As with determining a file's ownership and group ownership, you can use the ls command to find out a file's permissions. The ls command, without any options, does not present very much useful information. But with the -l option, the command lets you see the permissions of a file. The output will look something like this:

```
[jmoeller@fedora16test ~]$ ls -l
total 36
drwxr-xr-x. 2 jmoeller jmoeller 4096 Nov 12 18:32 Desktop
drwxr-xr-x. 2 jmoeller jmoeller 4096 Nov 12 18:32 Documents
drwxr-xr-x. 2 jmoeller jmoeller 4096 Nov 12 18:32 Downloads
drwxr-xr-x. 2 jmoeller jmoeller 4096 Nov 12 18:32 Music
drwxr-xr-x. 2 jmoeller jmoeller 4096 Nov 12 18:32 Pictures
drwxr-xr-x. 2 jmoeller jmoeller 4096 Nov 12 18:32 Public
drwxr-xr-x. 2 jmoeller jmoeller 4096 Nov 12 18:32 Templates
-rw-rw-r--. 1 jmoeller jmoeller  181 Feb  9 19:57 test
-rw-rw-r--. 1 jmoeller jmoeller    0 Feb  9 18:53
drwxr-xr-x. 2 jmoeller jmoeller 4096 Nov 12 18:32 Videos
```

The permissions for the various files and directories are on the left side of the command's output – the seemingly random strings of r, w, and x characters. However, once you know what these characters mean, you can determine the permissions of a file at a glance.

The "-" character at the start of the listing indicates that the entry belongs to a file; if there's a "d" there instead, it belongs to a directory. As for the rest of the characters, r stands for Read, w stands for Write, and x stands for Execute. It might seem strange that the permissions are repeated three times, but it's not – remember that permissions are assigned to three different classes of user for each file. Take, for example, a permissions string that looks like this:

rwxrwxrwx camalas users

The first "rwx" group means that the file's owner (in this case, camalas) has the Read, Write, and Execute permissions. The second "rwx" means that the owning group (the users group) has the Read, Write, and Execute permissions, while the third means that everyone else also gets the Read, Write, and Execute permissions. To put it succinctly, everyone on the Linux system can Read, Write, and Execute that file.

Let's look at a different example:

-rwxrw-r-- 6 camalas users 4096 Dec 9 14:56 test.txt

In this example, we see that the owner of test.txt, the user account for camalas, gets the Read, Write, and Execute permissions. The group gets Read and Write, while everyone else only gets the Read permission.

What if the permissions conflict? Like, what if camalas has only Read permission to a file, but everyone else gets Read, Write, and Execute? In that case, camalas gets the entire set of Read, Write, and Execute permissions to the file. Permissions are "additive" - that means your permissions to a file are "added up" from the owner's permissions, the group permissions, and everyone other account's permissions. Even if you're not the owner of a file, or a member of the owning group, you can still view and alter a file if the final class of permissions has the Read and Write permissions.

CHANGING PERMISSIONS

So, how to change a file's permissions? This is done using the chmod command. (In Linux, a file's total permissions are called the "mode", so chmod stands for "change mode.") As usual, if you want to change a file's permissions, you need to be root or using sudo to borrow root's powers. The chmod command uses both alphabetical and numerical shorthand to represent the various permissions.

If you want to add the Execute permission to the owning user of a file, you would use chmod like this (note that in chmod, u stands for the owning user, and x stands for the Execute permissions):

chmod u+x test.txt

If you wanted to add the Execute permission to the owning group, which is represented with a g in chmod:

chmod g+x test.txt

Finally, if you wanted to add the execute permission for everyone on the system:

chmod o+x test.txt

(In chmod, o stands for Others, or all other accounts on the system.)

And to do all three at once:

chmod u+x,g+x,o+x test.txt

Alternately, to remove permissions, replace the + sign with the - symbol:

chmod u-x,g-x,o-x test.txt

All this typing is rather cumbersome, isn't it? Fortunately, file permissions in Linux have numeric codes - mathematical shorthand, if you will. The Read permission is assigned the number 4, Write the number 2, and Execute the number 1. Having no permissions is represented by a zero. You then calculate the permissions by adding up the numbers. Take this example:

-rwxrw-r-- 6 camalas users 4096 Dec 9 14:56 test.txt

The file's owner (camalas) receives the Read, Write, and Execute permissions, the owning group gets Read and Write, and everyone else gets Read. Since Read is 4, Write is 2, and Execute is 1, camalas gets a permission of 7. The owning group of users gets a total of 6, and everyone else gets 4. The permissions can then be expression numerically as this:

764

And so to set the permissions as 764, you would use the chmod command:

chmod 764 test.txt

Much easier to simply type 764, isn't it?

Or to give the owning user full permissions, while denying permissions to both the owning group and everyone else, use this command:

chmod 700 test.txt

Needless to say, this makes using the chmod command far less cumbersome. As another example, let us say you wanted to change the permissions of test.txt so that the owner can read, write and execute, the owning group can read and execute, and everyone on the system who is neither the owner nor the owning user can read the file but not write or execute. You would use this chmod command:

chmod 754 test.txt

The chmod command, like chown and the file manipulation commands, also works with wildcards and the recursive -R switch. If, for example, you wanted to set permissions of 745 for the /test directory, you would use this command:

chmod -R 754 /test

With the chmod command, you can quickly and easily manage permissions from the command line on your Linux system.

15

CONCEPT: TEXT FILES

A great many of the files you will encounter on any Linux system are text files. In this chapter we'll explain what text files are and why it is important that you know how to use them.

WHAT IS A TEXT FILE?

A TEXT FILE IS, as the name suggests, a file that contains plain text and nothing more. If you view the file from the command line, you will be able to read the text.

Many files on your Linux system are a bit more complex. A .doc file, for instance, is the Microsoft Word document format. It contains a Word document formatted in digital form. The Word program takes that digital information and converts it into a text-based form for the end user. If you tried to open a Word file without using Word (or LibreOffice or another program that can read Word formats) you would only see gibberish.

There are other kinds of files that contain readable text, but also a great deal of other codes and formatting. An HTML file, a file for a web page, is a good example of this kind of file. If you view an HTML file from a command line, you'll be able to read it, but the actual text will include a great deal of HTML code. The file is, after all, designed to be displayed in a web browser, with the HTML code telling the web browser exactly how the file should look.

A plain text file, by contrast, contains no additional formatting and no code. Just plain, unadorned text. In a world of intricately formatted documents and feature-rich websites, a plain text file might seem useless, but text files are an important part of any Linux system.

CONFIGURATION FILES

TEXT FILES ARE significant because almost all of the configuration files, the files that control how the Linux operating system and its system services behave, are plain text files.

We've already discussed the /etc/passwd and /etc/group files, the two files that that control user accounts and groups in Linux. Both of those files are plain text files. Using certain commands, you could view them from the command line. You could even edit them, though editing those particular files is a bad idea (which is why we have the useradd, usermodd, and groupadd commands).

However, despite the restrictions on /etc/passwd and /etc/group, you change most of the other system settings on Linux by editing text files. If you're running a GUI interface, you can change system settings via a graphical interface, and very often server systems offer web interfaces for their programs. In the end, those programs are only graphical tools that edit the text files for you – you could just as easily edit those text files yourself.

We've already mentioned the hidden bash shell configuration

files in your home directory. If you use the ls -a command in your home directory, you will see several hidden files and folders. These are (almost) all text-based configuration files that govern your command-line session.

Most system-wide configuration files are found in the /etc directory. We've already mentioned the /etc/passwd, /etc/shadow, /etc/group, and /etc/gshadow files, all of which control user accounts and groups, and all of which are located in /etc. You can find the configuration files for the vast majority of your system's services and applications stored in /etc – either the files will be in /etc itself, or in a subdirectory within /etc.

User applications generally do not store their configuration files in /etc – they tend to wind up in /usr/lib. However, the majority of system configuration files and system service configuration files are plain text files stored in /etc. This makes knowing how to view plain text files, and how to edit them, important.

In the next to chapters, we'll show you how to do just that.

16

COMMANDS FOR VIEWING TEXT FILES

In this chapter, we'll show you how to view text files from the command line. Linux comes with numerous commands for viewing text files. All of them display text, but they each work a little differently, and some work better in certain situations than others. By the end of this chapter, you'll be familiar with the most common ones.

CREATING A BLANK FILE

YOU MAY NEED from time to time to create a blank file into which you will later dump some text. It's common to do this if you need to analyze the output from a command or a utility in greater detail, or to save it for later reference. Using the special output redirection greater than sign character (>) you can move the output from a command into a file. For instance, to redirect the output from the ls -l command to a file named data, the command would look like this:

```
ls -l >> data
```

You might want to first create the file, though. To create a completely blank file at the command line, use the touch command:

touch data

This will create a file named data in the current working directory.

VIEWING TEXT FILES WITH THE CAT COMMAND

THE BASIC COMMAND for viewing a text file is the cat command – it displays the contents of the file on the screen, and if it's too long to fit on the screen, the command simply scrolls down until it reaches the end of the file. To view a text file named document.txt in the current working directory, simply use the cat command like this:

cat document.txt

The output will display the contents of document.txt.

You can also use the cat command with absolute, rather than relative, file paths. If the current working directory is your home folder, and you want to view the config.txt file in the /etc directory, you would use this command:

cat /etc/config.txt

This will display the contents of the config.txt file.

Very often the contents of a text file are too large to fit on a single screen, and will go scrolling off the top before you can read it all. This is where the more command comes in handy. You can pipe the output from the cat command to the more command, which will then let you view the information one screen at a time. For example, if you wanted to view the contents of document.txt one screen at a time:

cat document.txt | more

This will display document.txt, pausing with every screen of information (you can press any key to continue).

You can also search the output of cat for any specific term by piping it to the grep command. The grep command lets you search a stream of text output for specific terms. Let's say you wanted to search document.txt for any instance of the word 'cheese.' To do so, you would pipe the output from cat to grep:

cat document.txt | grep cheese

This will list every instance of the word 'cheese' in the document.txt file.

The cat command also has a few useful switches. If you want the output from cat to appear in numbered lines, you would use the –b switch:

cat –b document.txt

The cat command, combined with the more and grep commands, lets you quickly view files from the command line, and find desired search terms within the file.

VIEW THE BEGINNING OF A TEXT FILE

THE CAT FILE displays the entire contents of a text file, but sometimes you might want to see only the beginning of the file. To do so, you can use the head command, which by default only displays the first five lines of a text file. For instance, to view the first five lines of a text file named test.txt in the current working directory, you would use this command:

head test.txt

The output will display the first five lines of the file.

The head command also works with absolute file paths. To view the first five lines of a the test.txt file located at /backup/test.txt, use this command:

head /backup/test.txt

The head command also allows you to choose how many lines you want displayed. Let us say you wanted to only view the first three lines of a the test.txt file:

head -3 test.txt

With that option, you can use head to view as many or as few lines from the beginning of a text file as you like.

VIEW THE END OF A TEXT FILE

IF THE HEAD lets you view the beginning of a text file, the tail command does the exact opposite – it displays the final five lines of a text file. For instance, to view the final five lines of a text file named test.txt in the current working directory, you would use this command:

tail test.txt

The output will display the final five lines of the file.

The tail command also works with absolute file paths. To view the final five lines of a the test.txt file located at /backup/test.txt, use this command:

tail /backup/test.txt

The tail command also allows you to choose how many lines you want displayed. Let us say you wanted to only view the final three lines of a the test.txt file:

tail -3 test.txt

With that option, you can use tail to view as many or as few lines from the end of a text file as you like.

SCROLLING THROUGH A TEXT FILE

THE CAT, head, and less commands all have one thing in common – they display the contents of a text file, and then quit. Even if you pipe the output of cat to the more command, you can scroll down one screen at a time – but you can't jump back to view a screen of previously shown text. If you're using a terminal emulator application, you can scroll back up, but if you're at the CLI interface, the text is gone.

This where the less command comes in handy. To turn to our example file from earlier, test.txt, this is the command you would use to view it with less:

less test.txt

Upon displaying the file, less pauses after the first full screen. By pressing the down arrow key, you can scroll down one line at a time. You can move down an entire screen at a time by pressing the Page Down key.

Unlike the other commands to view text files, less allows you to move back up and return to text that has already been displayed. Pressing the up arrow key allows you to move back one line at a time. Pressing the Page Up key moves up one screen.

You can search for terms within less by hitting the forward slash (/) key. After you hit the forward slash key, a prompt will appear at the bottom of the screen. Type your search term, and hit Enter. The less command will highlight each instance of the term within the output, and you can jump from instance to instance by hitting the n key.

Once you're done viewing the file, you can return to the command line by hitting the q key.

The cat, head, tail, and less commands will let you view text files, but they won't let you make any changes to those files. We'll show you how to edit text files in the next chapter.

17

EDITING TEXT FILES

The previous chapter showed you how to view text files from the command line. This chapter will explain how to edit text files from the command line. To do so, you use a program called a "text editor" that operates from the CLI. Usually, when you think of a text editor, a word processor like Microsoft Word or LibreOffice Writer comes to mind. An actual "text editor" is far more bare-bones, and devoted exclusively to editing text files.

There are a number of different text editors available, all with their own pros and cons. In this chapter, we will focus on a text editor called vi, which is included in almost all Linux distributions. The vi editor is also fairly easy for new users to learn, which is why we will discuss it here.

WHAT IS VI?

THE VI EDITOR has been around for a long time. It grew out of a series of early UNIX text editors in the 1960s and the 1970s, with the first version developed by programmer Bill Joy in the late 1970s. Later, vi was praised for its small memory footprint, which meant it could be used on systems with low amounts of available memory.

The variant of vi included with most versions of Linux is actually vim, which stands for "vi Improved." Vim was developed by a programmer named Bram Moolenaar in the early 1990s. Today, vim is generally the most popular variant of vi.

However, the vi command is still symbolically linked to vim, so typing vi into a Linux command line will launch vim.

LAUNCHING VI

LAUNCHING vi is simply a matter of typing the command:

vi

This will take you to an empty file, and you can create text files from here.

However, you'll often use vi to view or edit an existing text file. If you wanted to use vi to open a file named test.txt:

vi test.txt

The vi editor will then launch with the test.txt file open.

You can also use vi to create new files. For instance, if you wanted to use vi to create a text file named newfile.txt:

vi newfile.txt

However, note that the file isn't actually written to disk (and therefore doesn't exist permanently) until you save it. We'll discuss how to do that in the next section.

VI MODES

Vi is not like a graphical word processor, in that you can't simply sit down, fire it up, and start typing. Vi has different modes of operation, each of which has a different function. If you think of vi as a Swiss army knife, then the modes are the different kinds of blades and attachments within the knife.

Vi has four modes:

-Normal mode. Vi starts in default mode, and you can delete text from within normal mode.

-Command mode. This lets you save your text file.

-Insert mode. This lets you type fresh text into your file.

-Replace mode. This is similar to Insert mode, but you can overtype text, which you can't do in Insert mode.

Let's take a closer look at each of the modes.

NORMAL MODE

Vi starts in Normal mode. Normal mode lets you navigate around the text file by moving the cursor with the up and down arrow keys, along with the Page Up and Page Down keys. Pressing the HOME key takes you to the beginning of the file, and pressing END moves you to the end. You can also use Normal mode to delete text within the file, from single characters all the way up to entire lines. Normal mode also offers a search feature, allowing you to quickly locate specific search terms within the file.

First, let's take a look at the commands to delete text within Normal mode.

-Pressing dw deletes the word immediately after the cursor. This command also deletes the space following the word.

-Pressing de also deletes the word immediately following the cursor. Unlike dw, however, it does not delete the space following the word.

-Pressing d$ deletes from the cursor to the end of the current line.

-Pressing dd deletes the entire current line.

-Pressing p inserts the text from the last deletion after the current location of cursor. Essentially, in conjunction with the deletion commands, this lets you cut and paste within vi.

-Pressing u undoes the last action. Trust me, if you've made a mistake while editing a text file with vi, this can definitely come in handy.

-Pressing CTRL-G (the CTRL and the G keys at the same time) displays a status bar at the bottom of the screen. The status bar will display the name of the file, whether or not it has been modified since the last time it was saved, which line and column the cursor currently occupies, and how far into the file you are as a percentage.

-Pressing / (forward slash) activates the search function. Type a search term and hit enter, and vi will take you to the first instance of the search term. You can then hit the N key to jump to the next instance of the term.

Once you've deleted items from your text file, you may want to save the changes. We'll show you how to do that next in Command mode.

COMMAND MODE

NORMAL MODE LETS you navigate around a text file and delete parts of it, but Command mode lets you save your file. Command mode also lets you quit vi, with or without saving your text file first.

To access Command mode, you must first be in Normal mode. Press the colon key (the : key), and you'll be moved to Command mode. You'll know you're in Command mode when you see a : prompt at the bottom of your screen.

Let's look at the commands you can use while in Command mode (note that you must hit the Enter key after typing these commands):

-Pressing the w key writes the current text file to disk. This basically means that the file is saved. If you used vi to create a new file, it is written to disk. If you're editing a file, any changes you made to the file are written to the disk. So only use the w command if you're absolutely sure you want to save your changes.

-Pressing the q key quits vi and returns you to the command prompt. Note that the q command will only work if you haven't made any changes to the text file, or if you've made changes but haven't yet saved them to disk.

-Pressing the q! keys forces vi to quit, even if you haven't yet written your current text file to disk. This generally gets used when you've made serious mistakes editing a text file and want to start over from scratch. Remember, any changes you made to the text file are discarded, so only use q! if you don't want to save your changes.

-Pressing wq, as you might expect, combines the w and q commands. This command writes your file to disk, and then exits vi.

-Typing exit and then hitting the Enter key does the same thing as wq.

INSERT MODE

INSERT MODE, as the name suggests, lets you insert text into the text file. Access Insert mode by pressing the INSERT key while in Normal mode. You can also enter Insert mode by pressing the I key or the S key while in Normal mode.

Insert mode works much like most people expect a word processor to work – you type, and the text appears on the screen. You can also use the arrow keys and Page Up and Page Down to navigate around the file, even while in Insert mode.

Text is inserted to the right of the cursor. Note that Insert mode doesn't overwrite text when you type. Any text that already exists is pushed to the right as you add new text.

To return to Normal mode from Insert mode, press the ESC key.

Note that you can't enter Command mode from within Insert mode, since Insert mode interprets the colon character (:) as simply another text character and adds it to the file.

REPLACE MODE

THE FINAL MODE OF VI, Replace mode, works almost exactly like Insert mode, but with one key difference. Replace mode overwrites existing text as you type.

To access Replace mode, press the INSERT key while you are in Insert mode. This will shift you to Replace mode. As you type, you'll notice that preexisting text is overwritten, rather than moved to the right.

Once you've finished working in Replace mode, you can return to Normal mode by hitting the ESC key.

Note that you can't enter Command mode from within Replace mode, since Replace mode interprets the colon character (:) as simply another text character and adds it to the file.

WHY LEARN VI?

COMPARED to graphical word processors like LibreOffice or AbiWord, vi seems hideously complicated. Why learn it?

Because, as we mentioned above, you'll need it when working from the command line in an environment with no GUI. If you want to use Linux from the command line, you'll need to learn vi (or another command line editor). Besides, vi is actually pretty easy to learn. After some practice, you'll be editing text files with ease.

18

CONCEPT: NETWORKING

The days of the stand-alone home computer are long gone. Modern servers and workstations are almost always connected to the Internet, or at least a local area network. To effectively use a Linux system from the command line, you'll need at least a basic knowledge of TCP/IP computer networking, along with the network commands available from the prompt. In this chapter, we'll give you a basic overview of networking.

WHAT IS AN IP ADDRESS?

First, however, we should discuss IP addresses. Before we explain the Linux command prompt's networking commands, it's a good idea to have a firm grasp of the basics of IP addressing.

The letters "IP" stand for Internet Protocol, and the Internet Protocol is part of the TCP/IP (Transmission Control Protocol/Internet Protocol) suite, a group of related protocols that lay down the rules for how computers communicate over networks, both over

LANs and the larger Internet. An IP address, therefore, is a (theoretically) unique address assigned to a computer. It's a bit like a street address - it lets other computers send traffic to and receive traffic from your system. An IP address consists of four groups of numbers separated by dots:

192.168.1.1

These numbers are actually the numerical form of a binary number. None of these numbers can be higher than 255. (While watching a detective TV show, it's occasionally hilarious to see the police track a criminal using a ludicrously implausible IP address, like 689.34.385.339.)

The dominant version of the IP protocol is Version 4, commonly referred to as IPv4. Under IPv4, there are only 4.6 billion IP addresses available, and the available IP addresses ran out in late 2011. (IPv6, which has many more available addresses, will eventually replace IPv4, but for now, IPv4 remains dominant.) There are obviously far more computers, phones, routers, switches, and other networked devices in the world than 4.6 billion, so how do all these devices receive IP addresses?

The answer is a "private IP address." Certain blocks of IP addresses have been reserved for use in private networks. These blocks, using a technology called Network Address Translation (NAT), are then "translated" to public IP addresses. This has extended the lifetime of the available IPv4 address space for decades. The ranges of the reserved private addresses are:

10.0.0.0 - 10.255.255.255

172.16.0.0 - 172.31.255.255

192.168.0.0 - 192.168.255.255

Odds are, your computer has an IP address in one of those ranges as part of a private network (even if it's just a private network generated by your wireless router).

SUBNET MASKS

IP ADDRESS ALSO HAVE A "SUBNET MASK". A subnet mask defines which parts of the IP address designate the network, and which part designates the individual computer. Let's say the IP address of 192.168.1.1 from above has a subnet mask like this:

255.255.255.0

That means the 192.168.1 part of the address indicates the network, while the final 1 indicates the computer.

IP addresses usually include a "broadcast" address. Any traffic sent to the broadcast address is directed to every single computer in the local network. A broadcast address has a "255" as its final number, so a computer with a 192.168.1.1 address will have a broadcast address of 192.168.1.255.

DEFAULT GATEWAY

LASTLY, IP addresses usually (but not always) come with a "default gateway". The default gateway is the address you computer sends traffic destined for anywhere outside the local network segment. Like, say you want to visit Google with your web browser. Your computer recognizes that Google isn't on the 192.168.1.* network, and so forwards the request to the default gateway, which then sends the traffic on to Google. (This is a simplification, but adequate for our purposes.)

DNS

THE FINAL IMPORTANT piece of basic networking is the Domain Name System, or DNS.

As we mentioned above, every website on the Internet has, in theory, its own unique web address. Remembering all of those IP addresses can prove impractical. A few people might have keen enough memories to remember that microsoft.com has an IP address of 206.46.232.182 (at least at the time of this writing), but most of us cannot.

DNS provides an easier way to find websites. DNS maps each address to a domain name. When you type in a web address, such as microsoft.com or distrowatch.org, your request goes to a DNS server. The DNS server checks its database of domain name mappings (called "zone data") and responds with the IP address assigned to that domain name. Your computer is then able to contact the website and begin downloading data.

DNS is integral to the smooth functioning of most computer networks. If your computer is suffering connection problems, DNS is one of the first things to check (other than making sure the network cable is plugged in, of course).

In the next chapter, we'll show you the Linux commands that deal with basic networking.

19

NETWORKING COMMANDS

The last chapter explained the basics of networking. In this chapter, we'll show you how to perform networking tasks from the command line. You'll learn how to find your computer's IP address and MAC address, how to test connectivity, and how to find if your computer's DNS server is functioning properly.

FINDING YOUR IP ADDRESS

IT'S ACTUALLY a lot easier to find your IP address from the command line than from the GUI in most Linux distributions. Simply type this command at the prompt:

```
ifconfig
```

The ifconfig command will generate an output that looks something like this:

```
$ ifconfig
eth0      Link encap:Ethernet  HWaddr 08:00:27:9c:c4:83
          inet addr:192.168.2.11  Bcast:192.168.2.255  Mask:255.255.255.0
          inet6 addr: fe80::a00:27ff:fe9c:c483/64 Scope:Link
          UP BROADCAST RUNNING MULTICAST  MTU:1500  Metric:1
          RX packets:49 errors:0 dropped:0 overruns:0 frame:0
          TX packets:94 errors:0 dropped:0 overruns:0 carrier:0
          collisions:0 txqueuelen:1000
          RX bytes:19518 (19.5 KB)  TX bytes:12871 (12.8 KB)
```

There's quite a bit of information here, but most of it is useful. The "etho" refers to the first Ethernet connection on your system. The "indet addr" displays your system's IP address, while "Mask" shows the subnet mask. "HWaddr" shows your Ethernet adapter's MAC (Media Access Control) address, which is (theoretically) unique to each adapter. (Some wireless networks require you to supply your MAC address before allowing your system to connect.)

You can pipe the output from the ifconfig command to grep to quickly find the specific item you want. Let's say you just want to find the IP address:

ifconfig | grep inet

This time, the output will look like this, and you can quickly pick your IP address from the mix:

```
$ ifconfig | grep inet
          inet addr:192.168.2.11  Bcast:192.168.2.255  Mask:255.255.255.0
          inet6 addr: fe80::a00:27ff:fe9c:c483/64 Scope:Link
          inet addr:127.0.0.1  Mask:255.0.0.0
          inet6 addr: ::1/128 Scope:Host
```

Likewise, you can quickly find your computer's MAC address with ifconfig and grep. (A MAC address stands for Media Access Control, and is a hard-wired address that is theoretically unique to each network adapter.)

ifconfig | grep HWaddr

You can easily pick out your MAC address from the output:

```
$ ifconfig | grep HWaddr
eth0      Link encap:Ethernet  HWaddr 08:00:27:9c:c4:83
```

FINDING YOUR DEFAULT GATEWAY

THE IFCONFIG COMMAND generates lots of useful output, but it does not include the address of your default gateway. To do so, you can use the ip command with the route option:

ip route

The command will generate an output that looks something like this:

```
default via 192.168.2.1 dev p2p1  proto static
192.168.2.0/24 dev p2p1  proto kernel  scope link
```

Look for the line that begins "default via." Your system's default gateway will be the IP address after those two words. In this example screenshot, the default gateway would be 192.168.2.1.

OBTAINING AN IP ADDRESS

MOST OF THE TIME, your Linux system will receive an IP address from a DHCP server.

DHCP stands for "Dynamic Host Configuration Protocol", and is a kind of server that assigns an IP address to a client computer when it connects to the DHCP server's network. Configuring each individual computer with an IP address by hand is time-consuming and error prone; if one or more computers receive the same IP address, they won't be able to connect due to the address conflict. DHCP takes the guesswork out of the process. When a computer plugs into a Ethernet network (or connects to a wireless network), it sends out a request for a DHCP server. If there's a DHCP server on the network, it supplies the computer with an IP address, and the DHCP server ensures that no two computers receive the same IP address.

Most of the time, when you plug into a new network, you'll receive a DHCP address automatically. If you don't, however, try this command:

dhclient

Note that you will need to run dhclient as root, whether logged in as root or borrowing root powers via sudo.

This will manually tell your system to search for a DHCP server and accept an address. If this command doesn't work, that means either the network you're using doesn't have a functioning DHCP server, or that something is blocking the DHCP server from communicating with your computer.

RELEASING AN IP ADDRESS

DHCP addresses come with a "lease", which means the amount of time your computer is allowed to keep the address. Usually, your computer will contact the DHCP server at the halfway point before the lease expires, and ask for permission to keep it, which the DHCP server almost always grants.

Sometimes you want to release a leased IP address early - like if your network's DHCP server changes, or if your computer refuses to release its IP address. You can do this with the dhclient command and the -r option:

dhclient -r

This releases your IP address, and you can then use dhclient to contact a DHCP server to receive a new one.

TESTING YOUR DNS SERVER

As we mentioned above, if your DNS server stops responding, you can have all sorts of network problems. When it's working right, you will forget that it's there. But it goes bad, everything grinds to a halt. Without it, you can't type addresses into the address bar of your web browser, or use any network resources that depend upon DNS.

To diagnose your DNS server, you can use a command called nslookup. The nslookup command provides the IP address of your current DNS server, and also runs a test lookup against the server to see if it is functioning. For example, if you wanted to determine if your DNS server was operational, you would do so with this command:

nslookup distrowatch.org

The output would look something like this:

```
Server:         192.168.2.1
Address:        192.168.2.1#53

Non-authoritative answer:
Name:   distrowatch.org
Address: 66.180.174.35
```

The first two lines list the DNS server your computer is using. In this case, we are using a DNS server with an IP address of 192.168.2.1. The second group of lines shows the result of the lookup we performed – the IP address of distrowatch.org. In this case, you can see that distrowatch.org has an IP address of 66.180.174.35. Since nslookup successfully returned an IP address for distrowatch.org, you know that your DNS server is functioning correctly.

TESTING CONNECTIVITY

IF YOU'VE EVER HAD to repair a computer (or even used one, for that matter), you know that network and Internet connection problems are a depressingly common occurrence. Fortunately, the command line's tools are useful for obtaining valuable information about any network problems.

The first and most commonly used network-diagnostic command

is the ping command. The "ping" command stands for "Packet Internet Groper", and it relies upon the ICMP protocol, the Internet Control Message Protocol. Basically, the PING command sends a packet to a network destination, such as a server or another PC, and if the destination is active, it sends back an acknowledgment packet. For instance, if you wanted to ping the PC at the IP address of 192.168.1.2, you would use this command:

ping 10.10.11.254

If 10.10.11.254 is up, the ping command will generate an output like this:

```
64 bytes from 10.10.11.254: icmp_req=1 ttl=255 time=1.26 ms
64 bytes from 10.10.11.254: icmp_req=2 ttl=255 time=1.25 ms
64 bytes from 10.10.11.254: icmp_req=3 ttl=255 time=0.680 ms
```

The ping command will continuing pinging the target until you hit CTRL-C to cancel the command.

The command also has a few command options you can use to alter its performance. The -c option tells ping to send a set number of packets and then stop. For example, if you only wanted to ping 192.168.1.2 six times, you would use this command:

ping -c 6 192.168.1.2

Note that the ping command will not always work, even if a remote destination has an active network connection. This is because some operating systems by default blockICMP traffic. However, enough devices do permit ICMP traffic to make PING a useful diagnostic tool.

TRACING PATHS

SOMETIMES YOUR NETWORK connection is active, but you can't access a particular destination. In this case, the traceroute command becomes useful. The traceroute command maps the network route from your computer to the destination, and sends every device in the path an ICMP packet. You can then see which devices in the path are not responding, and then hopefully determine a solution. For instance, to trace the path to 10.10.11.254, you would use traceroute like this:

traceroute 10.10.11.254

The output will look something like this, with one line for each "hop" (another device in the chain) to the destination:

```
[jmoeller@fedora16test ~]$ traceroute 10.10.11.254
traceroute to 10.10.11.254 (10.10.11.254), 30 hops max, 60 byte packets
 1  10.10.11.254 (10.10.11.254)  1.086 ms * *
```

The traceroute command by default runs up to thirty hops.

20

CONCEPT: DISKS AND REMOVABLE MEDIA

Most of our commands so far have focused on working with the main hard drive in your Linux system. However, there are different kind of disks you can use with your Linux system – removable media, like external hard drives, USB flash drives, optical disks, and others. In this chapter we'll explain how the Linux operating system interacts with removable media, and in the following chapter we'll show you some commands for working with removable media.

NO DRIVE LETTERS

IF YOU'RE USED to working in a Windows environment, there is one important fact to keep in mind when working with disks on a Linux system – Linux does not use letters to represent separate drives!

In Windows, every disk or drive connected to the computer is assigned a drive letter. The computer's main hard drive usually gets the letter C. In the old days of floppy disks, the floppy drive got the letter A. The first optical drive on the system gets the letter D or E, generally. As you plug additional drives into a Windows computer, the Windows operating system assigns them the next available letter in the alphabet.

Linux uses a different system called mount points, which we'll describe in the next section.

MOUNTED DRIVES

Earlier in this book, we compared the Linux filesystem to a tree, with the root directory as the trunk, and the various subdirectories as branches shooting off from that main trunk. With Windows, each drive letter is in essence its own tree – the C drive gets its own root directory, as does the D drive and so on. When you plug a USB drive or insert an optical disk into a Linux system, the computer does not create a new filesystem tree. Instead, what Linux does is essentially graft a new branch onto the filesystem tree for the removable drive.

When a removable drive connects, Linux creates a "mount point" for that drive. A mount point is a virtual directory created in the filesystem for a removable drive. That directory usually receives the name of the drive's volume label. So if you plug a flash drive with a volume label of USBDRIVE into your computer, Linux will create a directory named USBDRIVE for it. The USBDRIVE directory will correspond to the top directory on the flash drive, and you can move files to and copy files from it just like any other directory.

Where do these mount points end up in the filesystem? Most Linux distributions have a /media directory to store mounted external drives. To return to our previous example, the USBDRIVE flash drive would show up in the filesystem as /media/USBDRIVE. For GUI interfaces, an icon representing the removable drive will show up on the desktop, but that icon will lead to the removable drive's mount point in /media.

Some Linux distributions use the /mnt directory, and a few even use /Volumes, but most distributions use the /media directory for mount points.

DIFFERENT KINDS OF FILESYSTEMS

WHETHER OR NOT YOU can read or write to removable media depends upon the kind of filesystem installed on the drive.

We've already described a filesystem as means of organizing information on a disk, and noted that "the filesystem" is often a shorthand term of describing the hard drives in a Linux system. While a filesystem may be a method of organizing information on a disk, there are nonetheless different kinds of filesystems. We'll take a look at a few of them in this section.

Hard drives in a Linux system are most commonly formatted with the ext3 or ext4 filesystems. Both ext3 and ext4 are journaled filesystems, which mean they keep track of changes before writing them to disk, which helps prevent errors. The ext3 filesystem supports volumes up to 32 terabytes in size, with a maximum size limit of two terabytes for individual files. The ext4 filesystem supports volumes up to one exibyte in size and files of up to 16 tebibytes in size.

If you plug an external hard drive into your Linux system, it might come formatted with the NTFS file system. NTFS is the default file system used with Windows computers. Most Linux distributions have the ability to read files on an NTFS volume, but not to write to them. If you plug an NTFS volume into your Linux computer, you'll probably be able to copy files from it, but you won't be able to write files to it.

The other major filesystem you are likely to encounter is the FAT file system. An older filesystem, FAT is generally not used on internal hard drives any longer, since its maximum possible size is four terabytes, and it does not support any files larger than four gigabytes.

However, both Linux and Windows can read from and write to FAT disks. So it is common for smaller external hard drives and USB flash drives to be formatted with FAT.

DIFFERENT KINDS OF REMOVABLE MEDIA

YOU WILL ENCOUNTER a few different kinds of removable media.

At the start of the 2000s, floppy disks were still quite common. Most floppy disks were the hard plastic 3 ½ inch disks, and could store 1.44 megabytes of data. This is not a lot of data by modern standards, and floppy disks were comparatively fragile and small. They have since been superseded and made obsolete by external hard drives, USB flash drives, and Internet-based storage systems, and are no longer manufactured.

External hard drives have become commonly and comparatively cheap. They usually connect to your computer via a USB cable. They come formatted with NTFS or the FAT file system. You can reformat an NTFS hard drive to FAT, or to ext3 or ext4, though this means a Windows system will not be able to recognize the drive.

USB flash drives are the most common type of removable media – they're cheap, easy to transport, and can hold a large amount of document files (though not very many music or video files). They're almost always formatted with FAT, so you'll have no trouble reading and writing files to them from a Linux system.

In the next chapter, we'll demonstrate some commands for getting the most out of removable media.

21

COMMANDS FOR WORKING WITH REMOVABLE MEDIA

The primary purpose of removable media is to move files on and off your computer. Most of the commands you'll actually want to use with removable media – cp, rm, mkdir, and so forth – we've already covered in this book. There are a few commands specific to analyzing disks and space usage that will prove useful when dealing with removable media. In this chapter, we'll show you how to use those commands.

FINDING DISK SPACE USAGE

YOU WILL OFTEN HAVE to calculate how much empty space remains on your disks, whether your removable media or your system's main hard drive. Granted, for a home computer, monitoring hard drive space usage isn't as critical as it used to be. Modern hard drives are enormous, and getting larger all the time. That said, certain kinds of multimedia files, such as video files, can quickly fill up a hard disk, and USB flash drives often will possess only a few gigabytes of space.

Fortunately, there's a quick and easy way to calculate just how much space you have left on your disks with the df command. To use it, simply type df at the prompt:

df

The output will look something like this:

```
Filesystem     1K-blocks    Used Available Use% Mounted on
/dev/sda1        7739864 3723760   3622940  51% /
udev              247172       4    247168   1% /dev
tmpfs             101672    1000    100672   1% /run
```

As you can see, the df command displays the space usage for every single filesystem attached to your system. The main hard drive on your computer will show up as /dev/sda1, with additional drives (whether internal hard drives or USB disks) showing up as /dev/sdb1, /dev/sdc1, and so forth. Any additional partitions on the drive will appear as additional numerals – /dev/sda2 and /dev/sda3 for the additional partitions on the first hard drive.

By default, df displays the space usage for the drives in bytes. Unless you're quite good at mental math, this is a difficult way to figure out exactly how much space you are using. To get an easier display to read, use df with the -h option:

df -h

The output will now look like this:

```
Filesystem      Size  Used Avail Use% Mounted on
/dev/sda1       7.4G  3.6G  3.5G  51% /
udev            242M  4.0K  242M   1% /dev
tmpfs           100M 1004K   99M   1% /run
```

With the -h option, df displays the output in gigabytes or megabytes, depending upon how large the disk is and how much space is in use.

You can also use the df command to find out what type of filesystem you have on your disks. To do so, use df -h with the addition of the -T option:

df -th

This will include the type of filesystem in the output:

```
Filesystem     Type      Size  Used Avail Use% Mounted on
/dev/sda1      ext4      7.4G  3.6G  3.5G  51% /
udev           devtmpfs  242M  4.0K  242M   1% /dev
tmpfs          tmpfs     100M 1004K   99M   1% /run
```

You can see that df now indicates the type of filesystem underneath the “type” column.

FINDING THE SIZE OF FILES AND FOLDERS

Now that you know how to find the space usage on your system's drives, you might want to know what, exactly, is taking up all that space on your hard disk. You'll remember that the ls -l command lists the sizes of files in the current working directory, and ls -lh displays the file sizes in megabytes and gigabytes. However, ls -l or ls -lh does not display the size of directories (and all the files contained in the directory) in the outputs. Because directories are technically also files in the Linux filesystem, the ls -l command only displays the size of the file marking the directory's existence, which is always four kilobytes.

To find the exact size of a directory, and all its subdirectories, use the du command. In its default state, the du command displays the size of all the files in the current working directory. However, combined with the -s and -h options, du will display the entire size of the targeted directory. For instance, to calculate the size of the /var directory and all its subdirectories, you would use du with these options:

du -sh /var

Note that for du to work, you'll need permissions to the target files and directories, which might mean you need to run du as root or with root powers.

UNMOUNTING REMOVABLE DISKS

USUALLY, on modern Linux systems, you don't have to worry about mounting USB drives. If you plug in a USB drive, the system will usually detect it (even on a CLI-based system) and mount it automatically. Unmounting a drive isn't quite so simple. In the GUI, you can just right-click on the drive's icon and click Remove Safely, but that doesn't work in the command-line interface. Simply yanking the drive out of its USB port isn't an option, since that might result in data corruption.

The answer is the umount command. Note that the command is actually spelled "umount", not "unmount". (Mistaking the two is a common typographical mistake at the Linux command line.) To return to our example from the previous chapter, if you wanted to unmount the /media/USBDRIVE flash drive, you would use this command at the prompt:

umount /media/USBDRIVE

Note that you need root powers to unmount mounted filesystems.

The umount command can fail. If you have a file open on the USB drive, the umount command fill refuse to unmount the filesystem. Or if you have another terminal session open, and your working directory on that session is on the drive to be unmounted, the umount command will fail. Basically, if any files on the drive are open, you cannot unmount it. However, you can attempt to forcibly unmount it with the -f option:

umount -f /media/USBDRIVE

Note that if you do have files open on the drive, this might result in data loss.

22

CONCEPT: PROCESSES AND DAEMONS

As you've read about Linux, you might have come across people discussing "processes" and "daemons." In the next few chapters, we'll discuss both. In this chapter, we'll describe what processes and daemons actually are, and then in the following chapter, we'll show you a few commands for monitoring and controlling processes.

WHAT IS A PROCESS?

In Linux, a "process" doesn't refer to the specific procedure you use to do something, or a standardized method for carrying out a task. Rather, a process is defined as the active instance of a computer program. For example, if you ran the Firefox web browser, Firefox would be a running process. Once you exit Firefox and close your browsing sessions, the process would exit and disappear.

Every program running on your computer, therefore, is technically a process. If you have a GUI, the various programs that make up the graphical user interface are processes. If you run a terminal emulator application, that too is a process. The commands you execute from the command line, whether in the pure CLI environment or from a terminal emulator application, are also processes. Even a command like ls, which quickly displays its output and then ends, is a process, albeit a short-lived one.

However, there are more programs on your Linux system than those you personally execute, and we will discuss those next.

DAEMONS

People sometimes refer to processes as "daemons", but the two terms are not interchangeable. While all daemons are indeed processes, daemons are something else. The term "daemon" means a system process running in the background. A system process is a process controlled and launched by the operating system without any intervention from the person actually using the computer. The program that controls a Linux laptop's power settings, for example, is a daemon, as are server applications like a Web server or MySQL.

(The term "daemon" originated not with the demons of Christianity or Islam, but with Greek myth. In ancient Greek religion, the "daemons" were lesser spirits who carried out tasks the Greek gods did want to bother themselves with. In the early days of UNIX, the programmers adapted the term "daemon" for processes that the system would carry out and not bother the user about.)

Generally, command-line beginners tend not to use daemons too often. Once your skills are more advanced, and you start installing server applications, you'll start using and controlling daemons more.

PROCESS ID NUMBERS

You'll recall from the chapters on users and groups that every user has its own user ID (UID) number and that ever group has its own group ID (GID) number. Processes are no different. Every process that launches on a Linux system receives its own process ID (PID) number. Even a short-lived process like the ls command gets its own PID when it runs.

After a process runs, it (in theory) releases its process ID number back into the pool of available processes. The number is then available to be reassigned to any new process.

It is important to understand how PIDs work, since many of the commands that control processes use PIDs to identify the target of the command.

PARENT PROCESS ID NUMBERS

Every process also has a PPID – a parent process ID number.

What is a parent process? Every process on a Linux system is launched by another process. If you run a terminal emulator and launch a command from the application, the terminal emulator process launches the process for the command. Or if you launch a web browser application, and you open a PDF file through the browser, the browser process will launch the PDF reader application.

This relationship is called the parent process and the child process, similar to the relationship between a directory and a child directory. The parent process is the first process, and it launches the child process. Every process, therefore, has a PPID – the PID number of the process that launched it.

THE INIT PROCESS

Of course, if every process is launched either by a user or another process, then what is the first process?

The init process runs during the boot sequence, and launches all other processes on the system. It has a PID of 1, and is the first process launched during boot. Every system process is ultimately launched by init.

In the next chapter we'll show you a few commands for managing processes.

23

COMMANDS FOR MANAGING PROCESSES

In this chapter, we'll show you how to view running processes from the command line. We'll also show you how to shut down hung or frozen processes.

VIEWING RUNNING PROCESSES

One of the simplest methods of viewing the processes running on your system is the venerable ps command. To use the ps command, simply type it at the command prompt:

ps

However, without any options, ps is not terribly useful. The default output looks like this:

```
  PID TTY          TIME CMD
 2187 pts/0    00:00:00 bash
 2238 pts/0    00:00:00 ps
```

In its default state, ps only shows the currently running process – itself – and the shell you are currently using. Fortunately, you can make ps much more useful with the –e switch:

ps –e

The –e stands for Extended, and generates a listing of every process currently running on the system:

```
PID TTY          TIME CMD
  1 ?        00:00:00 init
  2 ?        00:00:00 kthreadd
  3 ?        00:00:00 ksoftirqd/0
  5 ?        00:00:00 kworker/u:0
  6 ?        00:00:00 migration/0
  7 ?        00:00:00 cpuset
```

However, this still only displays a limited amount of information, especially when compared to the top command (which we will discuss later in this chapter). You can tell the ps command to display more information with the –l switch:

ps -el

The output will look like this:

```
F S   UID   PID  PPID  C PRI  NI ADDR SZ WCHAN  TTY          TIME CMD
4 S     0     1     0  0  80   0 -   799 poll_s ?        00:00:00 init
1 S     0     2     0  0  80   0 -     0 kthrea ?        00:00:00 kthreadd
```

As with the ls command, the –l switch stands for Long Listing, and generates quite a bit more information. Specifically, you can see the UID (user ID) of the user who launched the process, the PPID (parent PID) of the parent process of a child process, the process's priority, and numerous other useful pieces of information. For even more information, use ps with the –f switch:

ps –elf

With these options, the output from ps will look like this:

```
F S UID          PID  PPID  C PRI  NI ADDR SZ WCHAN  STIME TTY          TIME CMD
4 S root           1     0  0  80   0 -   799 poll_s 09:11 ?        00:00:00 /sbin
1 S root           2     0  0  80   0 -     0 kthrea 09:11 ?        00:00:00 [kth]
```

In addition to the information offered by the –l switch, the addition of the –f switch also causes the ps command to display the date the process was launched and the location of its executable on the file system.

Even ps –elf is only a static snapshot of the processes currently running on the system. The running processes can change very quickly on a Linux system, and you might want to monitor them in real-time. That's where the top command comes in handy.

MONITOR PROCESSES WITH THE TOP COMMAND

THE TOP COMMAND has been around forever, and is included in practically every variant of Linux. It displays a variety of useful statistics, including the "top" users of CPU time (hence the command's name), the amount of memory running processes are using, the amount of virtual memory they are using, the process's priority, the process's owning user, and numerous other useful pieces of information. To launch top, simply type the command at the command prompt:

top

The command's output will look like this:

```
top - 13:26:48 up  4:15,  1 user,  load average: 0.00, 0.01, 0.05
Tasks: 116 total,   1 running, 115 sleeping,   0 stopped,   0 zombie
Cpu(s):  1.0%us,  0.7%sy,  0.0%ni, 98.3%id,  0.0%wa,  0.0%hi,  0.0%si,  0.0%st
Mem:   1026080k total,   628596k used,   397484k free,    97884k buffers
Swap:  1046524k total,        0k used,  1046524k free,   340772k cached

  PID USER      PR  NI  VIRT  RES  SHR S %CPU %MEM    TIME+  COMMAND
 2245 jmoeller  20   0 68204  14m 9.8m S  1.3  1.4   0:00.53 gnome-terminal
 1001 root      20   0 54128  31m 5012 S  1.0  3.1   0:07.76 Xorg
    1 root      20   0  3196 1808 1200 S  0.0  0.2   0:00.75 init
    2 root      20   0     0    0    0 S  0.0  0.0   0:00.00 kthreadd
    3 root      20   0     0    0    0 S  0.0  0.0   0:00.02 ksoftirqd/0
    5 root      20   0     0    0    0 S  0.0  0.0   0:00.36 kworker/u:0
    6 root      RT   0     0    0    0 S  0.0  0.0   0:00.00 migration/0
```

Once launched, top stays running, displaying statistics in real time. Processes are listed by their process ID number, along the name of the command that started the process.

There's a lot of information listed in top, so let's take a look at the available statistics:

PID: A process's process ID number.

USER: The process's owner. Processes owned by root are usually system processes, and should be left alone unless they cause problems.

PR: The process's priority, which determines how much attention the CPU will give this process over other processes. The lower the number, the higher the priority.

NI: The nice value of the process, which affects its priority.

VIRT: How much virtual memory the process is using.

RES: How much physical RAM the process is using, measured in kilobytes.

SHR: How much shared memory the process is using.

S: The current status of the process (zombied, sleeping, running, uninterruptedly sleeping, or traced).

%CPU: The percentage of the processor time used by the process.

%MEM: The percentage of physical RAM used by the process.

TIME+: How much processor time the process has used.

COMMAND: The name of the command that started the process.

Of course, you may want to sort the top display by a certain category from least to greatest. This is especially useful when you want to figure out what process is hogging the most CPU time or memory. Hitting the 't' key will sort the processes by CPU time. Hitting 'l' will sort by load average, and 'm' by memory info.

You can also use top to kill uncooperative processes. To do this, you'll need to make note of the process ID you want to kill. Hit the "k" key while top is running, and the utility will ask you to input a process ID number. Input the number, hit Enter, and then hit Enter again when it asks if you want to kill the process (top assumes you mean "yes" when you hit enter). The top utility will then kill the process.

Once you've finished with top, hit the q key to exit and return to the command line.

Most often, you'll use the top and ps commands to track down a process that's causing trouble. In the next section, we'll discuss how to terminate troublesome processes with the (appropriately named) kill command.

STOPPING PROCESSES

THE KILL COMMAND, as you might guess from the name, lets you kill running processes.

Technically, the kill command sends signals to running processes. However, this command is most commonly used to send the kill signal to a running process. In most cases, you will use this command to terminate a process that has frozen up, refuses to quit gracefully, or is otherwise misbehaving.

Note that to use kill, you need to know the PID (process identification number) of the process you wish to terminate. You can find out the PID using either the ps or the top commands (detailed previously in this chapter). Also, unless you launched a specific process, you will need root powers to kill it.

To use kill, simply type the command with the PID of the targeted process. For instance, to terminate a process with the PID of 2589:

kill 2589

Like other commands, kill can be modified with a number of options. Remember that kill operates by sending signals to the designated process. The kill command uses 64 separate signals, which you can list with the –l switch:

kill –l

Of those 64 signals, four of them are commonly used. You tell kill which signal to send by using the signal's number as a command option – for instance, kill- 9 2589.

The 15 signal is the default kill signal. 15 tells the targeted process to cease immediately, but only after shutting down gracefully. Since 15 is the default kill signal, you don't usually have specify it, but if you did, the command to send the 15 signal to PID 2589 would look like this:

kill -15 2589

Signal number 1 less drastic. It merely tells the process to restart, while keeping the same PID. To use signal number 1:

kill -1 2589

Signal number 2 sends the CTRL+C character combination to a process, which is commonly used to break out of a running process (the ping command, for instance). To send signal number 2 to a process:

kill -2 2589

Finally, the most drastic signal is number 9. Signal number 9 immediately kills a process, and does not give it the option to close gracefully or clean up after itself first. Generally, you should only use signal number 9 to terminate a process if nothing else has worked first:

kill -9 2589

Using ps, top, and kill, you can manage the running processes on your Linux system from the command line, and quickly respond to any problems.

24

CONCEPT: INSTALLING SOFTWARE

Linux's built-in suite of commands and utilities are quite useful, and many Linux distributions come with an array of powerful applications. However, no single distribution will come with every application you will conceivably need. At some point you'll have to install software on your Linux system. In this chapter we'll explain how install software on a Linux system works, and the next chapter will show you some commands for doing just that.

DIFFERENCES FROM WINDOWS

FOR PEOPLE USED to Microsoft Windows, it can be difficult to figure out how to install software on a Linux system. Usually, on a Windows machine, you download the installer file from the website, double-click on it, and follow the steps offered by the install wizard. Or you insert the CD or DVD into your optical drive and follow the AutoPlay prompt, or navigate to the DVD's root menu through Windows

Explorer and double-click on the SETUP.EXE file. One of the strengths of Microsoft Windows is the relative ease of installing software. (In fact, this ease of installing software on Windows is something of a weakness – consider how easy it is to get a virus on a Windows computer!)

Installing software on a Linux system used to be much harder than it is now, and the graphical installation tools have progressed greatly in the last few years. That said, it is still quite different than installing software on a Microsoft Windows system.

SOFTWARE PACKAGES

MOST SOFTWARE for Linux system comes in the form of a software "package". A software package is an installer that has been prepared to work with a specific Linux distribution. On any Linux distribution, it is in fact possible to compile your own executable files from the source code. (This is why when a software developer publicly posts his code, his application is known as "open source" - the source code is open.) Compiling your own software from source, however, doesn't always work if you don't know what you're doing. There's also the problem of "dependencies" other applications or services the software needs to install properly. Early Linux systems could encounter a condition known as "dependency hell", where you have to install dependency after dependency in order to get the original application you wanted up and running.

Prepackaged software takes a lot of the guesswork out of installing software on a Linux system. Usually, packaged software has been prepared to work with a specific distribution of Linux, and if you use a packager manager to install it, the application knows which dependencies it needs.

A "package manager" is an application designed specifically to install packaged software. There are two major packaging tools in

Linux, depending on whether or not your distribution is derived from Debian or Red Hat/Fedora, two of the most influential Linux distributions. Distributions derived from Red Hat/Fedora tend to use the RPM Package Manager, which is usually referred to as RPM. Distributions derived from Debian often used the Advanced Packaging Tool, usually abbreviated down to apt. Package managers are useful because they know what dependencies installed software needs, and will automatically install the dependencies with the chosen application.

SOFTWARE REPOSITORIES

PACKAGE MANAGERS usually work closely with a software repository.

Basically, a software repository is a server on the Internet where the packaged software for a specific distribution is stored. Repositories are usually maintained by the maker of a specific distribution. For instance, Canonical, the company behind Ubuntu, maintains the software repositories that Ubuntu uses. Red Hat, the company that creates Red Hat Linux and Fedora Linux, runs the repositories for both of those distributions.

Software repositories offer many advantages. Since the company behind the distribution maintains it, you can find software specifically tested and approved to work with your distribution. This means it's also easy to get updates for your installed software, since updates are added to repository after they are tested to work with a distribution. The downside to a repository is that you need Internet access in order to use it.

In the next chapter, we'll show how to install software from the command line.

25

COMMANDS FOR INSTALLING SOFTWARE

In this chapter, we'll explain how to install software from the command line. We will focus on the two major package managers – the Advanced Packaging Tool and RPM. We'll also show you how to install prepackaged software without using a software repository.

INSTALLING PREPACKAGED SOFTWARE ON A DEBIAN-BASED SYSTEM

Debian is one of the oldest and most respected Linux distributions, and numerous other distributions are based upon it – Ubuntu, Kubuntu, Xubuntu, Linux Mint, Knoppix, and Xandros are just a few of the Linux distributions derived from Debian. Software packaged for Debian-based distributions usually comes in the form of a .deb file. You can install the package in the GUI environment by double-clicking on it, but you can also install it from the command line.

Note that to install or uninstall any software, whether from the

GUI or from the CLI, you will need root-level powers.

To install .deb packages from the command line, you will use the dpkg command. If you have a prepackaged installer file called installer.deb, you can install it with this command:

dpkg -i installer.deb

Depending on the program, you may or may not have to follow an install wizard or enter additional information. Follow the prompts, and dpkg will install the software for you.

Likewise, you can also use dpkg to uninstall software with the -r option:

dpkg -r installer.deb

Note that in order to uninstall software via dpkg, you need to keep the original installer file, since that contains the instructions for uninstallation. Also, the dpkg command does not calculate any dependencies, and if you're missing a dependency the piece of software needs, the installation will fail.

INSTALLING SOFTWARE USING THE ADVANCED PACKAGING TOOL

Most Debian-derived distributions used the Advanced Packaging Tool for software management. The Advanced Packaging Tool has a number of graphical front ends – the Synaptic Package Manager, or the Ubuntu Software Center in Ubuntu. However, to install software from the command line with the Advancing Packaging Tool, you'll use its command-line utility, the apt command. The apt command works closely with the software repositories, and downloads its installed software from the repository.

For instance, to install the remote desktop connection application Remmina, you would use this command:

apt-get install remmina

Usually on a Debian-derived system you'll use apt with sudo to gain root powers for the installation. You'll also need to know the name of the package you want to install. After you enter the command, apt will check the repositories, list any dependencies Remmina might need, and list the total amount of disk space required. Press Y to continue, and apt will download and install Remmina for you.

To remove Remmina with apt, you can use this command:

apt-get remove remmina

Again follow the prompts, and apt will remove Remmina and all its dependencies from your system. Note that if one of Remmina's dependencies is needed by another application, apt will not remove it.

INSTALLING PREPACKAGED SOFTWARE USING THE RPM PACKAGE MANAGER

LIKE DEBIAN, Red Hat Linux (and its free version, Fedora) is one of the oldest and most respected Linux distributions, and numerous other distributions are based upon Fedora. Software packaged for Fedora-based distributions usually comes in the form of a .rpm file, since Red Hat, Fedora, and its derivatives all use the RPM Package Manager instead of the Advanced Packaging Tool. You can install the package in the GUI environment by double-clicking on it, but you can also install it from the command line.

To install .rpm packages from the command line, you will use the rpm command. If you have a prepackaged installer file called installer.rpm, you can install it with this command:

rpm -i installer.rpm

Depending on the program, you may or may not have to follow an install wizard or enter additional information. Follow the prompts, and rpm will install the software for you.

Likewise, you can also use rpm to uninstall software with the -e option:

rpm -e installer.rpm

Note that in order to uninstall software via rpm, you need to keep the original installer file, since that contains the instructions for uninstallation. Also, the rpm command does not calculate any dependencies, and if you're missing a dependency the piece of software needs, the installation will fail.

INSTALLING SOFTWARE WITH YELLOW DOG UPDATER

MOST FEDORA-DERIVED DISTRIBUTIONS used the Yellowdog Updater Modified (yum) for software management. Yellowdog has a number of advantages over the rpm command – it works closely with the repositories, and automatically calculates and downloads dependencies. Yellowdog uses the yum command-line utility to install software.

For instance, to install the remote desktop connection application Remmina, you would use this command:

yum install remmina

Usually on a Fedora-derived system you'll use yum as root to gain the necessary permissions for installation. You'll also need to know the name of the package you want to install. After you enter the command, yum will check the repositories, list any dependencies Remmina might need, and list the total amount of disk space required. Press Y to continue, and yum will download and install Remmina for you.

To remove Remmina with yum, you can use this command:

yum remove remmina

Again follow the prompts, and yum will remove Remmina and all its dependencies from your system. Note that if one of Remmina's dependencies is needed by another application, yum will not remove it.

26

GETTING HELP

The Linux command line is a complex environment. We've done our best to show you the basics, but we can't cover every single option and scenario here. And as you grow in experience and knowledge, you will want to perform more advanced tasks than covered in this book. Fortunately, Linux comes with a comprehensive built-in help system, and we'll show you how to use it in this chapter.

APROPOS

THE FIRST STEP in using the help system is the apropos command.

Sometimes you know what task you want to perform from the command line, but you don't know what command you want to use. This is when the apropos command comes in handy, since it lets you search by topic or keyword. For example, let's say you want to perform a task involving the ext3 filesystem, but you can't quite recall the precise command. You can do a search with the apropos command:

apropos ext3

The command will generate an output that looks like this:

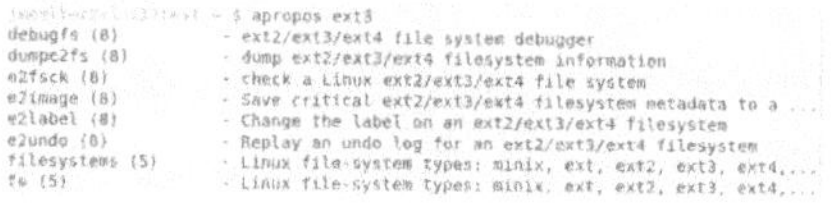

```
$ apropos ext3
debugfs (8)          - ext2/ext3/ext4 file system debugger
dumpe2fs (8)         - dump ext2/ext3/ext4 filesystem information
e2fsck (8)           - check a Linux ext2/ext3/ext4 file system
e2image (8)          - Save critical ext2/ext3/ext4 filesystem metadata to a ...
e2label (8)          - Change the label on an ext2/ext3/ext4 filesystem
e2undo (8)           - Replay an undo log for an ext2/ext3/ext4 filesystem
filesystems (5)      - Linux file-system types: minix, ext, ext2, ext3, ext4,...
fs (5)               - Linux file-system types: minix, ext, ext2, ext3, ext4,...
```

The output will list every command that has "ext3" in its description, allowing you to jog your memory.

Where does apropos get its information? The command draws its output from Linux's built-in manual pages, which we'll describe in the next section.

MAN PAGES

EVERY COMMAND in Linux has (at least in theory) an accompanying "manual page", commonly shortened to "man page." The man page is a text document describing the command, detailing its options, and providing examples for its use. Man pages were first included in UNIX in the early 1970s, and appear in almost every distribution of Linux.

How do you access the man page for a particular command? You do this using the man command, followed by the name of the command for which you wish to view the manual page. For instance, if you wanted to view the man page for the nslookup command, you would use this command:

man nslookup

This will open the man page for the nslookup command.

Once the man page is open, it acts quite a bit like the less command. You can use the up and down arrow keys to scroll through the text line by line, along with the PAGE UP and PAGE DOWN keys to move through a page at a time. Like the less command, the man utility supports searches. Hit the forward slash (/) key to type a search term, and the man utility will jump to the first instance of the term in the text. You can then hit the N key to jump from instance to instance.

Once you've finished reading the man page, you can exit by hitting the Q key to return to the prompt.

What information can you expect to find in a man page? The actual layout of a man page will vary from distribution to distribution, but there are usual several standard sections.

The first section is the title, which will list the name of the command.

The second section is the name, which will again display the name of command, and provide a short summary of its function.

The third section is the synopsis, which provides a brief description of the command's syntax.

The fourth is the description, which is the actual meat of the man page. In the description, you'll find the options and functions of the command listed in detail. There are usually examples of the command's usage here as well.

The final sections of the man page display the programmer of the command itself, an address for reporting bugs, copyright information, and other man pages that might be relevant.

Using the man utility, you can obtain useful information on any command on your Linux system.

ABOUT THE AUTHOR

Standing over six feet tall, Jonathan Moeller has the piercing blue eyes of a Conan of Cimmeria, the bronze-colored hair a Visigothic warrior-king, and the stern visage of a captain of men, none of which are useful in his career as a computer repairman, alas.

He has written the DEMONSOULED series of sword-and-sorcery novels, and continues to write THE GHOSTS sequence about assassin and spy Caina Amalas, the COMPUTER BEGINNER'S GUIDE series of computer books, and numerous other works.

Visit his website at:

http://www.jonathanmoeller.com

Visit his technology blog at:

http://www.computerbeginnersguides.com

Contact him at:

jmcontact@jonathanmoeller.com

www.ingramcontent.com/pod-product-compliance
Lightning Source LLC
Chambersburg PA
CBHW031242050326
40690CB00007B/923
9781718177079